T0360779

Allocation in Networks

Allocation in Networks

Jens Leth Hougaard

The MIT Press
Cambridge, Massachusetts
London, England

This book was set in Times by Westchester Publishing Services.

Library of Congress Cataloging-in-Publication Data

Names: Hougaard, Jens Leth, author.
Title: Allocation in networks / Jens Leth Hougaard ; foreword by Hervé Moulin.
Description: Cambridge, MA : MIT Press, [2018] | Includes bibliographical references and index.
Identifiers: LCCN 2018005104 | ISBN 9780262038645 (hardcover : alk. paper)
Subjects: LCSH: Resource allocation. | Cost—Mathematical models.
Classification: LCC T57.77 .H6798 2018 | DDC 338.8/7—dc23 LC record available at
    https://lccn.loc.gov/2018005104

To Yi and Julie

Contents

Foreword

The analysis of networks and graphs is a field of mathematics that is not much more than seventy years old. The breadth of its applications is astonishing, from physics and biology to the social sciences and "Internet science." In this volume, Prof. Jens L. Hougaard surveys a particularly active subset of this research, at the interface of operations research, economics, and computer science: it focuses on the thorny problems of resource allocation in the construction and exploitation of networks.

One of the iconic combinatorial optimization problems, that of finding the cheapest tree connecting users (on the nodes of the network) to a single source, doubles as a cost-sharing problem if nodes are occupied by users who need to connect with the source. Solving convincingly the latter question is one of the great successes of cooperative game theory in the past three decades. Chapters 2–4 take us in clear incremental steps from the fair solution of this computationally simple connectivity problem to the much more complicated task of allocating the cost of general networks under arbitrary connectivity demands of the users, and to the related problem of dividing the profits generated by a flow graph among the owners of its nodes.

In chapters 5 and 6 the network becomes the result of the decentralized choices of its individual users, exactly like social media and other communication networks. The noncooperative game of network formation is now at center stage; we learn, inter alia, about the social cost of such decentralized actions, measured by the influential concept of the "price of anarchy." Finally, the author calls on the methodology of mechanism design to align the incentives of the network users with economic efficiency.

The reader will enjoy the breezy style of exposition, entirely self-contained, that is long on carefully chosen examples to present a broad range of results and short on irrelevant technicalities. This approach reveals the underlying unity of models coming from significantly distinct academic literatures. Detailed references and numerous exercises complete this remarkable book, which will appeal to seasoned researchers no less than to beginner students.

Hervé Moulin

Preface

This book is about networks and economic design. It takes a regulator's viewpoint and focuses on the role played by allocation rules (that is, revenue and cost-sharing schemes) in sustaining, as well as creating, efficient network solutions. The text presents an up-to-date overview of models and results that are scattered across several strands of literature, primarily in the fields of economics, operations research, and computer science. The topic has been an integral part of my own research agenda for the past decade, so the book naturally includes much of my recent research.

The book is intended as teaching material for an advanced economics course at the graduate or PhD level, but it should be equally relevant, at least in parts, for students of operations research and computer science. It is also my hope that researchers in the field as well as practitioners in network regulation—and in the network industry—will find the book useful as a reference. To facilitate the reading, the text is accompanied by numerous illustrative examples, and to challenge the reader, the end of each chapter contains a set of exercises meant to highlight the concepts and methods involved as well as present the reader with further details of analysis.

Networks are also important in research. Much of the work presented in this book is the result of a longstanding collaboration with coauthors (and friends): Peter Østerdal, Juan D. Moreno-Ternero, Hervé Moulin, and Mich Tvede. I am most grateful to them for being a constant source of inspiration and for their feedback on this book project. In particular, I thank Hervé for writing the Foreword.

My research for this book has further benefited from joint work, and inspiring discussions with Parimal Bag, Gustavo Bergantiños, Jens Gudmundsson, Chiu Yu Ko, Dorte Kronborg, Richard Ma, Peter Bro Miltersen, Kurt Nielsen, Aleksandrs Smilgins, Rene van den Brink, and Jingyi Xue.

A special thanks to Frank Jensen for extensive and detailed comments on earlier drafts of this book.

Finally, thanks to the MIT Press team: in particular, to Emily Taber for her support and encouragement from the very first email.

Jens Leth Hougaard
Copenhagen, November 2017

Introduction

Networks, both in a physical and social sense, have always played an important role in human activities. This is emphasized by the recent boom in economic and social activities enabled by the Internet and mobile communication devices. Just by looking at the worth of companies like Google and Facebook, it is evident that the various kinds of positive synergies created by network activities are able to generate substantial user value. Networks can also facilitate coordination and information sharing among any type of physical devices that can be switched on and off the Internet. The far-reaching consequences of this type of *Internet of Things*[1] have only just begun to appear, and the potential economic implications are predicted to be huge. For instance, McKinsey&Co writes: "we estimate that the Internet of Things has a total potential economic impact of \$3.9 trillion to \$11.1 trillion per year in 2025. On the top end, the value of this impact—including consumer surplus—would be equivalent to about 11 percent of the world economy in 2025."[2]

In light of their importance, it is not surprising that networks have been intensively studied in a wide range of scientific fields, from physics to anthropology. The present book focuses on economic aspects of agents sharing a common network resource, typically in the form of a distribution system. The book asks fundamental questions, such as how agents ought to share costs and benefits from coordinated actions in various network structures, and how the choice of such allocation methods influence the network structure itself through different types of incentivizing effects. Thus, the book takes a normative approach, typically from the viewpoint of a benevolent

---

1. Defined as sensors and actuators connected by networks to computing systems.

2. McKinsey Global Institute, "The Internet of Things: Mapping the Value Beyond the Hype," https://www.mckinsey.com/ /media/McKinsey/Business%20Functions/McKinsey%20Digital/Our %20Insights/The%20Internet%20of%20Things%20The%20value%20of%20digitizing%20the% 20physical%20world/The-Internet-of-things-Mapping-the-value-beyond-the-hype.ashx,June2015.

social planner, or a regulator, aiming to ensure economically efficient network solutions sustained by distributional fairness.

The main focus is therefore not on social and economic network formation, where agents decide with whom to connect and the decisions shape both network structure and agents' benefits. This research agenda has been predominant in the economics literature for the past two decades.[3] We do study network formation, but only as far as it concerns allocation issues: for instance, when examining the relation between allocation mechanisms and their induced effects on the stability and efficiency of the resulting network structure.

We will mainly consider models and techniques from operations research in combination with tools from cooperative game theory and microeconomics, in order to systematically analyze fair allocation when agents share a common network.[4] A guiding research agenda since the 1970s has been to ask what type of allocation problems, in what type of networks, ensure the existence of stable allocations in the sense of the standalone core (i.e., allocations for which no coalition of agents have incentive to block socially efficient coordination). As witnessed by the coming chapters, this agenda has been very fruitful, applying solution concepts from cooperative game theory to allocation problems arising in various network models from combinatorial optimization.

However, practice may not always fit the model of cooperative game theory. It seems that many network structures are the result of an evolutionary process, in which the network evolves, more or less efficiently, over time. When users face networks in the short run, the latter can therefore be construed as fixed and often inefficient, and the value of alternative options for subgroups of agents are at best counterfactual and typically unknown. Recent research efforts therefore tend to move in the direction of analyzing model-specific allocation methods, which at times may be in conflict with the standalone core but have compelling and easy to interpret axiomatic foundations. The contrast between fair and reasonable allocation when agents take over (or gain access to) an existing network en bloc, and when the network is designed specifically to satisfy their needs (i.e., solving an underlying optimization problem) is a recurrent theme throughout this book.

Using the theory of mechanism design, we will also ask what kind of allocation mechanism a social planner (regulator) should apply with the aim of implementing an efficient network when the information between planner and agents is asymmetric: typically agents are informed about costs and benefits,

---

3. See, e.g., Goyal (2007), Jackson (2008).
4. See, e.g., Moulin (1988, 2004), Sharkey (1995), and Curiel (1997).

while the planner is not. This asymmetry in information leaves ample room for the agents to manipulate the planner's choice of network via misreporting. This seriously limits the range of useful allocation rules for the planner. The theory of mechanism design is somewhat abstract, focusing on general conditions for implementability. However, we will study simple game forms in the specific context of connection networks.

In a decentralized setting, where the planner's role is limited to defining the "rules of the game," we study agents' strategic reactions to the planner's choice of allocation mechanism. As mentioned above, the guiding research agenda in economics (since the mid-1990s) has focused on why networks are formed, whether they are stable, and whether stability is compatible with efficiency when the agents themselves can choose to form and maintain connections.[5] Results demonstrate that there is a tension between network stability and social optimality, which goes deeper than what should be expected at first glance from the mere presence of network externalities.

The economics literature is supplemented by a related research agenda in computer science.[6] In large networks it is often difficult, if not impossible, to obtain a centrally coordinated solution. A natural question is therefore when equilibria arising from individual optimization are good approximations of socially optimal outcomes. When quantifying the difference between selfish and coordinated actions, the central notions are the *price of anarchy* and the *price of stability*, that is, the ratio between welfare in the worst (respectively, best) equilibrium and welfare in social optimum. Quite surprisingly, these ratios are bounded for many problems, and we can start searching for conditions, such as the choice of allocation rule, under which the worst-case price of anarchy/stability is minimized.

Why Study Allocation in Networks?

Allocation issues that arise in networks are practical problems that require practical operational solutions. In practice, nonmarket allocation typically takes place through negotiation or bargaining procedures, or by using simple rules of thumb for reasons of transparency and computational ease. So what can a systematic theoretical investigation of allocation rules offer to practitioners who already solve these problems every day? The short answer seems at least twofold:

5. See, e.g., Jackson (2008).
6. See, e.g., Roughgarden (2005), Nizan et al. (2007).

• We need benchmarks for practical solutions in the form of theoretically compelling normative solutions. Welfare theory provides a number of allocation rules with proven desirable properties for large classes of allocation problems. Decision makers may not like them for all kinds of political or practical reasons, but at least the rules are available and can inspire and inform any potential discussion. Historical evolution with repeated practical interactions may lead to normatively sound allocation schemes, but why not try to get it right from the start? For regulatory purposes, the normative allocation rules are often directly applicable in various tailor-made versions.

• The need for theoretically sound allocation mechanisms also arises when allocation decisions are left to computational agents, as in many new electronic markets. For instance, when system resources are allocated in data flows with congestion, we obviously cannot rely on direct bargaining processes among users when literally thousands of decisions have to be made in split seconds and must be made continuously. Careful selection of allocation mechanisms is crucial, since this has a major impact on system efficiency and development, and thereby on user welfare both in the short and the long run.

In a centralized scenario, a benevolent social planner is able to enforce a socially optimal network solution and can justify such a solution by fair allocation of the resulting costs, or benefits, among network users. More importantly, however, fairness is often needed to sustain the socially optimal network, making sure that no subcoalition of users will have incentive to create parallel, or alternative, subnetworks.

The study of fair allocation can be traced back to the earliest human writings and has remained a central theme in several scientific fields, including economics.[7] Specific to the modern approach of welfare theory is that it builds on explicit criteria or properties (formulated as axioms) that characterize different notions of fairness. This makes it rather practical and "hands-on" compared to the subtle abstract approaches of political philosophy: for example, John Rawls's widely debated Difference Principle, stating that inequalities are only justified if they are to the greatest benefit of the least advantaged members of society.

Providing an axiomatic foundation for various types of allocation rules allows for a qualitative comparison of these rules. Such characterizations can help select the right allocation rule based on the desirability of the characterizing properties in the given context. For instance, the idea of fairness embodied

---

7. See, e.g., Young (1994), Roemer (1996), Hougaard (2009), and Thomson (2016).

in the standalone core conditions seem particularly compelling when designing an optimal network, knowing users' values for all possible network options.

Some of the allocation rules we will investigate are related to a derived cooperative game, and characterizing axioms (with their inherent notions of fairness) often relate directly to the derived game itself. Other rules are defined with respect to the network structure, and the characterizing axioms are typically directly related to the network structure as well. In particular, some rules have representations (and interpretations) in the form of step-by-step procedures that are closely related to solution algorithms for the underlying network-optimization problem. Thus, the axiomatic framework can also help determine what rules work well in what types of networks.

In a decentralized scenario, choosing an allocation mechanism is an important instrument for the planner. Typically, a tradeoff takes place between fairness and efficiency, in the sense that fair allocation rules may not induce formation of efficient networks, and conversely, ensuring that efficient networks are formed may imply unfair distribution of payoffs.

Working with formal definitions of allocation mechanisms in stylized models allows us to investigate how various forms of network externalities influence optimal network structure and allocation of payoffs. In practice a negative externality, such as congestion, often has a major impact on flows and distribution in networks. With our formal models, we can pinpoint these impacts. For instance, we can analyze how the choice of allocation method incentivizes agents and whether this leads to equilibrium behavior that comes suitably close to socially optimal behavior. Moreover, we can investigate the details of the trade-off between fairness and efficiency once we are able to characterize fairness via well-defined properties of the allocation rule. Since an understanding of the overall interaction between the choice of allocation rule and the resulting network structure is essential when managing large networks with autonomous agents, any scientific progress made in this area can potentially have a significant impact on social welfare.

The next section provides a few specific examples illustrating the importance of understanding how allocation mechanisms influence network efficiency and user welfare in practice.

Motivating Examples

To demonstrate that the relevance of understanding network allocation issues stretches far beyond the age of the Internet, let us start with a historical example dating back to the twelfth century.

The German Hansa

The German Hansa began as an association of north German merchants in the twelfth century, but it developed into an entire community of cities at its peak around the middle of the fourteenth century.[8] The purpose of the community was to obtain trading privileges by monopolizing markets and to protect and support its members in their commercial endeavors.

As a community the Hansa was remarkably stable and remained active for almost 500 years. It is worth noting here that the community was founded on *voluntary participation*. The Hansa can be seen as a set of merchants and towns forming a network of markets and routes for commodity transport. In the language of graphs, the set of nodes (or locations) is in this case the markets (or towns) where the trade takes place, and the edges connecting the nodes are the roads and seaways used for commodity transport.

It was often very costly for the Hansa to establish new trading routes and markets, simply because they were facing a hostile environment of brigands and piracy, making roads and seaways highly unsafe. For instance, the pagan inhabitants of Finland and the Baltic countries made trade with Russia very risky. So around the year 1200, Bishop Albert led a crusade into the Baltic countries co-financed by the Hansa. In particular, members based on Gotland, who stood to benefit from opening up commercial opportunities in Russia, contributed to this crusade by equipping hundreds of crusaders and providing for their transportation. Likewise, in 1241 the towns of Hamburg and Lübeck agreed on sharing the costs of keeping interconnecting roads free from brigands.

Once transportation routes were secured and market privileges obtained, these new "edges and nodes" of the Hanseatic network were public goods for the community. But this is not the same as saying that all members would benefit equally. The distribution of benefits depended on individual trading patterns (i.e., connectivity demands). It seems that no one has studied in detail the cost-sharing schemes used by the Hansa, but it appears that since it was founded on voluntary participation, no member was forced to subsidize activities for which they did not benefit directly. For example, the Hansa was at war with Denmark between 1367 and 1369, when Denmark asked for peace. The Westphalian towns (including Cologne) did not contribute financially or support this war, but they were not excluded by the Hansa

---

8. A standard reference on the German Hansa is Dollinger (1970). For a network interpretation, see Hougaard and Tvede (2015).

simply because it was known that these towns mainly traded with England and the Netherlands and therefore had no commercial interests in the war with Denmark.

However, the Hansa did not allow members to free ride. For example, the Norwegian king restricted Hanseatic privileges in 1241, and the Hansa responded with a blockade. One member, Bremen, did not participate in the blockade and was excluded by the Hansa. The Hansa must have seen the act by Bremen as an attempt to free ride: if the blockade failed, Bremen was free to continue trading with Norway; if it was successful, Bremen would obtain Hanseatic privileges.

The stability of the Hansa indicates that the community was capable of finding fair ways to share the costs and benefits of their cooperation subject to network constraints and individual connectivity demands. When the Hansa finally dissolved in the seventeenth century, it was not due to internal disagreements on allocation issues, but rather as a consequence of a changing political landscape and the Protestant reformation.

Jumping ahead in time, the next example will show that the choice of allocation rule may influence the ability of a team to locate hidden "treasure."

The DARPA Network Challenge

The U.S. Defense Advanced Research Projects Agency (DARPA) announced a social network mobilization experiment on October 29, 2009, (the fortieth anniversary of what is considered the birth of the Internet) to "explore the roles the Internet and social networking play in the timely communication, wide-area team-building, and urgent mobilization required to solve broad-scope, time-critical problems."[9] In short, ten moored red weather balloons were placed at ten undisclosed locations in the (continental) United States. A prize of $40,000 would be given to the first participant who submitted correct geographical data concerning the location of all ten balloons within the contest period. This task was believed to be impossible by conventional intelligence-gathering methods and required an explicit Web-based social mobilization strategy.

The challenge was won by the Massachusetts Institute of Technology (MIT) team, who managed to locate all ten balloons in a bit less than 9 hours. The MIT team realized that the key to success lies in incentivizing agents (here balloon finders) in the right way. They decided to use the $40,000 prize money as the basis for a financial incentive structure in which not only the finder

---

9. See http://archive.darpa.mil/networkchallenge/ and the paper by Pickard et al. (2011).

would be paid, but also the person who recruited the finder, and the person who recruited the person who recruited the finder, and so forth. Specifically, they managed to recruit almost 4,400 people by promising $2,000 to the first person who send in the correct coordinates of a balloon; $1,000 to the person who recruited a finder; $500 to the recruiter of this recruiter, and so on. Thus, the diffusion of information about tasks was obtained by initiating a recruitment network in which individuals were given incentives to both act to solve the task itself and to recruit others for its solution.

The victory of the MIT team demonstrates that the choice of allocation rule has important incentivizing effects on those involved. For instance, if the prize money were split among balloon finders only, the incentive for initiating the tree-structured recruitment process would be gone, and the powerful tool of information diffusion would be lost. In this book, we will examine a normative foundation for the type of half-split transfer rule in hierarchical network structures used by the MIT team, showing that this type of rule rests on compelling fairness properties as well.

The final example illustrates a key role of networks in an economy, which is to deliver resources to multiple users with different demands and cost characteristics. Such distribution systems are recognized as natural monopolies, as in the case of electricity transmission.

Regulating Transmission

In transmission systems, generators and consumers are the two main types of actors involved, and the costs they must share relate to maintenance, planning, and operation of the system. These costs must be covered by network charges that are subject to various forms of regulation.

With recent reforms in many countries, the power industry has become unbundled in the sense that noncompetitive segments, such as network core facilities, are separated from potentially competitive segments involving production, supply, and maintenance. Transmission open access is one of the competitive elements in these reforms, and with shared use of the grid comes the problem of allocating the associated costs (e.g., fixed transmission costs) in a fair and reasonable way that reflects the actual usage of the network by those involved.

Due to the importance and practical nature of the problem, there is a large literature suggesting, and analyzing, several different allocation methods (with funny names like the *Postage Stamp* and the Z_{bus} method).[10] Since both

10. See, e.g. Pan et al. (2000) and Brunekreeft et al. (2005).

generators and consumers are likely to react to the way network charges are set (i.e., costs are allocated), the structure of these charges will heavily influence the efficient use and development of the network.

In line with the general approach in the coming chapters of this book, we shall (briefly) consider some ideal requirements of cost-allocation methods from a regulator's viewpoint. The literature generally agrees that the regulator ought to make sure that the applied cost-allocation method ensures the technical quality of the transmission service, is transparent and easy to regulate, and also covers costs. But on top of that (and more interesting from an economic viewpoint) it is important that the charges are fair (e.g., in the sense of no cross-subsidization) and lead to efficient short-run use of the network as well as provide the right incentives for new transmission investment.

Obviously these are very demanding requirements for any specific method to satisfy in full, and in practice some compromises will have to be made. Also, the choice of suitable method is likely to depend on the structure of the actual transmission system. For instance, if the network has a tree structure (often called a *radial* structure) it is easy to identify flows from generator to consumer, while this is much more complicated in grids (often called *mesh* structures). The fact that network topology plays a role when establishing what is fair and reasonable is emphasized throughout the coming chapters and even guides the overall organization of the models that will be presented.

Plan of the Book

In terms of organization, this book uses network topology as a way to categorize and sort the many different models scattered across the literature.

After recalling some basic definitions and results from graph theory and cooperative game theory in chapter 1, we start out by looking at models where the network has a *tree* structure in chapter 2. Trees appear in many models from operations research and economics, and they typically capture situations where a set of agents (broadly interpreted) are connected to a single source (supplier). In the absence of externalities—and if connections are costly and reliable—efficiency results in tree-structured networks. In practice, trees appear in the form of organizational hierarchies, client-server computer systems (where individual "stars," consisting of a central server with a set of peripheral clients, are connected), water-distribution systems (such as irrigation ditches), and the like. In a tree there is a unique path between any two nodes (e.g., an agent and the source), making it relatively simple to determine individual network usage and thereby to find fair and reasonable allocations. It becomes considerably more complicated when adding, for instance, public nodes (Steiner nodes) or congestion externalities.

In chapter 3, we review a series of models from combinatorial optimization, known under the common name of *routing problems*, where the structure of the network is given by a *cycle*. For example, a salesperson has to visit a number of clients and return to office, or a postman has to deliver mail to a number of streets before returning to the post office. The total cost of the tour has to be covered by the agents involved. But cycles may also result from redundancy or demand for increased connectivity, as in electrical source loops. At first glance, there seems to be little difference between connecting a set of agents to a source and going from a source to each of the agents. But the fact that the "salesman" has to return to the source (creating the cycle) actually makes the two problems radically different, since it is no longer evident how the individual agent connects to the source (should we consider the path coming *to* the agent from the source, or the path going *from* the agent to the source?). This challenge is not an easy one for fair allocation. The chapter presents several ways to deal with this problem.

Chapter 4 examines general network structures in which tree and cycle structures are combined in various ways. In particular, we consider allocation in inefficient networks, where some connections are redundant. In contrast to tree structures, "flows" and connectivity in general networks are considerably more complicated. Thus, determining agent-specific usage of (or service to) the network becomes much more complex. The chapter will look at different models and approaches to handle this challenge. This includes an analysis of allocation issues in flow problems, for instance, when agents control certain links and have to cooperate in order to obtain a maximal flow, as well as a generalization of Myerson's graph restricted game model (presented in chapter 1), where the societal value of every network configuration can be assessed.

Chapter 5 considers a decentralized scenario where networks are operated or formed by competitive autonomous agents. We study how the use of given allocation rules influences strategic behavior, which in turn may affect the structure of the network itself when agents have full discretion to form and maintain connections in the network. We will consider models of selfish routing and study the ability of allocation mechanisms to induce desirable equilibrium behavior in the sense of approximating social optimality. We also will consider the tension between efficiency and stability of networks: insisting on certain critical fairness properties when distributing the common value of network formation turns out to exclude the possibility of creating efficient networks that are robust to individual obstructions. The question is whether we can compensate agents so as to maintain connections that are not in their own interest, but are desirable from a social perspective. We also will study how efficient networks may be obtained through a suitably designed sequential bargaining process among network users, and how certain direct mechanisms can elicit truthful revelation of users' private valuations of network service at the cost of a potential efficiency loss.

In chapter 6, we come back to a centralized scenario, but this time with asymmetric information between a social planner and the agents for whom the network is designed: agents are fully informed (e.g., about connection demands and costs); the planner is not. We study how the planner can design a game form (mechanism) that implements a socially efficient network under such circumstances. In particular, we study two (cost-minimization) models from chapters 2 and 4: the minimum-cost spanning tree model and its generalization, the minimum-cost connection network model, under different game forms. Moreover, when agents have limited willingness to pay for connectivity, the challenge is to design a welfare-maximizing network that may exclude certain agents simply because the net social benefit of adding them to the network is negative. Although it becomes more difficult to obtain efficient implementation, it is still possible to find mechanisms that work (in theory).[11]

---

11. In chapters 5 and 6, the analysis takes for granted that the reader is familiar with basic notions from noncooperative game theory, such as Nash equilibrium, strong Nash equilibrium, and subgame perfect Nash equilibrium. If not, the reader is referred to one of the many excellent books on game theory (e.g., Fudenberg and Tirole 1991, or the more recent Maschler et al. 2013).

References

Brunekreeft, G., K. Neuhoff, and D. Newbery (2005), "Electricity Transmission: An Overview of the Current Debate," *Utilities Policy* 13: 73–93.

Curiel, I. (1997), *Cooperative Game Theory and Applications: Cooperative Games Arising from Combinatorial Optimization Problems.* Kluwer Academic Publishers.

Dollinger, P. (1970), *The German Hansa.* Macmillan.

Fudenberg, D, and J. Tirole (1991), *Game Theory.* Cambridge, MA: MIT Press.

Goyal, S. (2007), *Connections: An Introduction to the Economics of Networks.* Princeton, NJ: Princeton University Press.

Jackson, M. O. (2008), *Social and Economic Networks.* Princeton, NJ: Princeton University Press.

Hougaard, J. L. (2009), *An Introduction to Allocation Rules.* New York: Springer.

Hougaard J. L. and M. Tvede (2015), "Minimum Cost Connection Networks: Truth-Telling and Implementation," *Journal of Economic Theory* 157: 76–99.

Maschler, M., E. Solan, and S. Zamir (2013), *Game Theory.* Cambridge: Cambridge University Press.

Moulin, H. (1988), *Axioms of Cooperative Decision Making.* Cambridge, MA: Cambridge University Press.

Moulin, H. (2004), *Fair Division and Collective Welfare.* Cambridge, MA: MIT Press.

Nizan, N., T. Roughgarden, E. Tardos, and V. Vazirani (2007), *Algorithmic Game Theory.* Cambridge: Cambridge University Press.

Pan, J., Y. Teklu, S. Rahman, and K. Jun (2000), "Review of Usage-Based Transmission Cost Allocation Methods under Open Access," *IEEE Transaction on Power Systems* 15: 1218–1224.

Pickard, G., W. Pan, I. Rahwan, M. Cebrian, R. Crane, A. Madan, and A. Pentland (2011), "Time Critical Social Mobilization," *Science* 334: 509–512.

Roemer, J. E. (1996), *Theories of Distributive Justice*, Cambridge, MA: Harvard University Press.

Roughgarden, T. (2005), *Selfish Routing and the Price of Anarchy*, Cambridge; MA: MIT Press.

Sharkey, W. W. (1995), "Network Models in Economics." In Ball et al., (eds) *Handbooks in Operations Research and Management Science,* 713–765.

Thomson, W. (2016), "Fair Allocation." In M. D. Adler and M. Fleurbaey (eds.), *The Oxford Handbook of Well-Being and Public Policy.* Oxford: Oxford University Press.

Young, H. P. (1994), *Equity.* Princeton, NJ: Princeton University Press.

1 Some Basics

Before analyzing allocation problems in various network structures, it is convenient to review a few basic notions and results from graph theory and cooperative game theory as well as some interconnections between graphs and cooperative games.

1.1 Graphs

The mathematical concept of a *graph* is both a natural and convenient way to represent a network. We therefore review a series of basic definitions and results from graph theory (see, e.g., Bollobas, 1979, 2002).

1.1.1 Notation and Basic Definitions

A *graph* is a pair of sets $G = (N, E)$, with N being a set of nodes (vertices) and E a set of edges (connections, links). Typically, we assume that both N and E are finite sets: $N = \{1, \ldots, n\}$ and $E = \{1, \ldots, m\}$.

In models where the set of nodes, N, is identified with the set of agents, the set of edges, E, represents connections between these agents. Connections can be physical, for instance, in the form of wires, cables, pipes, or roads. Examples include energy transmission networks, water distribution systems, and computer networks (as in client-server and peer-to-peer networks). But connections can also be virtual, for instance, in the form of alliances, communication, relationships, or hierarchies. Examples include wireless cellular networks and various forms of social and business networks.

In models where the set of edges, E, is identified with the set of agents, the set of nodes, N, represents locations between which we can traffic goods or services to the agents.

When the graph is *undirected*, each edge $e \in E$ joins an unordered pair of nodes $i, j \in N$, so $ij = ji$. Many networks are undirected: for instance, roads can be used to drive both ways, and agents can communicate with each other or form mutual trading agreements.

For edge $e = ij$, the nodes i and j are called *endnodes*. Two edges are said to be *adjacent* if they have exactly one common endnode. Edges having the same endnodes are called *parallel*. Edges of the form $e = ii$ are called *loops*. Graphs with no parallel edges and no loops are called *simple* graphs.

If $E = E^N$ is the set of all pairs of nodes in N, we say that the graph $G^N = (N, E^N)$ is *complete*. A complete graph has $\frac{n(n-1)}{2}$ edges (i.e., pairs of nodes, since each one of n nodes can be paired with any of the remaining $(n-1)$ nodes and then adjusting for the fact that each node is counted twice).

For notational convenience, we write $G + e$ ($G - e$) for the graph that consists of the graph G plus (minus) the edge e.

Edges $e = ij$ in a graph may have associated *weights* c_{ij}, which can be interpreted as values, for instance, the capacity, cost, or benefit associated with the link between nodes i and j. In this case G is called a *weighted graph*.

A graph G can also be represented by its $n \times n$ *adjacency matrix* $A = A(G) = (a_{ij})$, where a_{ij} is the number of edges with i and j as endnodes. For a simple graph, $a_{ij} = 1$ if $ij \in E(G)$ and $a_{ij} = 0$ otherwise. When the graph is undirected, the adjacency matrix is symmetric around its main diagonal, since $a_{ij} = a_{ji}$. If the simple graph is weighted, the *weighted adjacency matrix* is defined by $a_{ij} = c_{ij}$ if $ij \in E(G)$ and $a_{ij} = 0$ otherwise.

Two graphs $G = (N, E)$ and $G' = (N', E')$ are *isomorphic* if there exists a bijection $\lambda : N \to N'$ such that $ij \in E$ if and only if $\lambda(i)\lambda(j) \in E'$. *G and G' are isomorphic if and only if, for some ordering of the nodes, their adjacency matrices are identical.*

The *degree* $\delta(i)$ of a node $i \in N$ is the number of edges with i as endnode. For a graph $G = (N, E)$, we have

$$\sum_{i=1}^{n} \delta(i) = 2m, \tag{1.1}$$

since when summing the degrees over all nodes, each of the m edges is counted twice.

Remark   (1.1) can be used to show that in any graph there is an even number of nodes with odd degree (indeed, divide all nodes into two groups—those with even and those with odd degree. The sum of degrees in the even group is even and by the observation above, the total sum is even too, so the sum of the odd group is also even. Thus, since the sum of degrees in the odd group is the sum of odd numbers, there must be an even number of nodes in that group).

In a simple graph the maximal number of edges adjacent to any node i is $n - 1$. Thus, a straightforward way to express the *degree of centrality* of node i is to use the measure

$$\frac{\delta(i)}{n-1} \in [0, 1]. \tag{1.2}$$

The two polar cases are that i is connected to all other nodes, yielding degree of centrality 1, and that i is a singleton node, yielding degree of centrality 0.[1] When agents are identified by nodes, the degree of centrality may be relevant for the way values are allocated. Some rules will favor a central agent; others will punish a central agent, depending on the specific context.

Two nodes i and j are *connected* in $G = (N, E)$ if there exists a sequence of edges (a *path*) $i_1 i_2, i_2 i_3, \dots, i_{h-1} i_h$ such that $i_k i_{k+1} \in E(G)$ for $1 \leq k \leq h - 1$, where $i = i_1$ and $j = i_h$. A graph $G = (N, E)$ is said to be *connected* if i and j are connected in G for all $i, j \in N$. The complete graph G^N is said to be *fully connected*. An edge e, in a connected graph $G = (N, E)$, *disconnects* G if the graph $G - e$ is disconnected. A disconnecting edge is often called a *bridge*.

In models where agents are identified by nodes, a connected network offers connectivity between any pair of agents, either directly or indirectly, via one or multiple paths. For a given agent, the set of paths that deliver the desired connectivity represents the agent's potential network usage.

Remark   In a graph $G = (N, E)$ with two distinct nodes i and j, the minimal number of edges disconnecting i from j is equal to the maximal number of edge-disjoint paths connecting i and j (Menger's Theorem).

If $N' \subset N$ and $E' \subset E$, the graph $G' = (N', E')$ is said to be a *subgraph* of $G = (N, E)$. In particular, the subgraph $G' \subset G$ is a *component* of G if it is maximally connected. Indeed, *a graph is connected if and only if it has exactly one component.*

Example 1.1 (Illustrating the basic concepts)   Let $N = \{1, 2, 3, 4\}$ be a set of four nodes, and consider the simple graph $G = \{(14), (23)\}$ (shown below), consisting of the two edges connecting endnodes 1 and 4, and endnodes 2 and 3, respectively. Clearly this graph is disconnected. Adding, say, the edge 12 will make the graph $G' = G + \{12\}$ connected (i.e., there is a path in G' connecting any pair of nodes in N). As such, the edge 12 is a bridge in G' connecting the two components (each consisting of a single edge) 14 and 23.

---

1. The literature defines several alternative centrality measures, which will not be reviewed here.

Removing the edge 14 in the graph G results in two singleton nodes 1 and 4 as well as the component (edge) 23. Note that in general removing a bridge from a graph results in two components (or singleton nodes). Clearly, G and G' are *not* isomorphic. As an example of a graph isomorphic to G' we could select $G'' = G + \{34\}$.

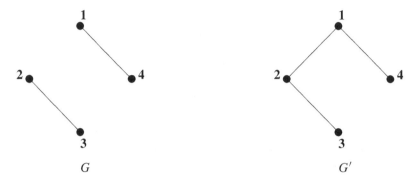

The degree of centrality is $1/3$ for all nodes in G, while it is $2/3$ for nodes 1 and 2 (and $1/3$ for nodes 3 and 4) in G'.

Graphs G and G' can be represented by their adjacency matrices:

$$A(G) = \begin{pmatrix} 0 & 0 & 0 & 1 \\ 0 & 0 & 1 & 0 \\ 0 & 1 & 0 & 0 \\ 1 & 0 & 0 & 0 \end{pmatrix}$$

$$A(G') = \begin{pmatrix} 0 & 1 & 0 & 1 \\ 1 & 0 & 1 & 0 \\ 0 & 1 & 0 & 0 \\ 1 & 0 & 0 & 0 \end{pmatrix}.$$

1.1.2   Cycles

A sequence of edges is called a *cycle* if it starts and ends with the same node (i.e., if $i_1 i_2, i_2 i_3, \ldots, i_{h-1} i_h, i_h i_1$).

A cycle for which all nodes and edges are visited once (except for the start, and end, node) is called a *circuit*.

A cycle in a graph G is called an *Euler cycle* if it includes all edges and all nodes of G. *A graph has an Euler cycle if and only if it is connected and every node has even degree* (the Euler/Hierholzer Theorem).

Fleury's algorithm can be used to find an Euler cycle in a connected graph where every node has an even degree. The algorithm basically says: start at an arbitrary node and move along nondisconnecting edges (unless a disconnecting edge is the only possibility); delete the edges as you move along (as well as the node left behind in the case of disconnecting edges).[2]

Example 1.2 (Finding an Euler cycle)   The graph below provides an example with possible Euler cycles.

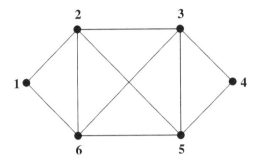

Using Fleury's algorithm, we can start at node 1 and take the following tour: go to node 2 (delete edge 12); go to node 3 (delete edge 23); go to node 4 (delete edge 34); go to node 5 (delete edge 45 and node 4); go to node 3 (delete edge 35); go to node 6 (delete edge 36 and node 3); go to node 5 (delete edge 56); go to node 2 (delete edge 25 and node 5); go to node 6 (delete edge 26 and node 2); go to node 1 (delete edge 16 and node 6). We have thus identified an Euler cycle.

In the case of a weighted (connected) graph, the problem of finding a minimal-weight Euler cycle is also known as the *Chinese postman* problem (Kwan, 1962). Starting from the post office, a postman has to visit each street (edge) at least once before returning to the post office. There is a delivery cost (weight) for each edge (street), and the postman has to plan the tour such that the total delivery cost is minimized. In a cost-sharing context, we return to this problem in chapter 3.

A cycle in a graph G is called a *Hamiltonian cycle* if it contains each node exactly once, except for the starting and ending node (which appears twice). Thus, a Hamiltonian cycle is a circuit.

Clearly, each node has to have at least degree 2 if there is to be any hope of constructing a Hamiltonian cycle in a given graph. *For $n > 3$ a sufficient condition for the existence of a Hamiltonian cycle is that each node has at least degree $n/2$* (Dirac's Theorem). However, it is also clear that this is not a necessary condition.

---

2. For alternative algorithms, see, e.g., Eiselt et al. (1995).

Remark For $n \geq 3$ the complete graph G^N has $(n-1)!/2$ possible Hamiltonian cycles. Indeed, starting at a given node i, there are $n-1$ possible connections to the next node j; from node j there are $n-2$ possible connections to the next node, and so forth, until only one edge leads back to node i. That is, there are $(n-1)!$ possible directed cycles and therefore $(n-1)!/2$ undirected cycles.

For $n \geq 3$ the complete graph G^N is decomposable into edge-disjoint Hamiltonian cycles if and only if n is odd. There are $\frac{1}{2}(n-1)$ such edge-disjoint Hamiltonian cycles, since in a cycle every node has degree 2, and for each node the set of $n-1$ remaining nodes can be partitioned into $\frac{1}{2}(n-1)$ pairs.

Example 1.3 (Edge-disjoint Hamiltonian cycles) Consider the complete graph G^N with $N = \{1, \ldots, 5\}$. As mentioned in the remark above, the complete graph can be decomposed into $\frac{1}{2}(5-1) = 2$ edge-disjoint Hamiltonian cycles in this case. These are illustrated below.

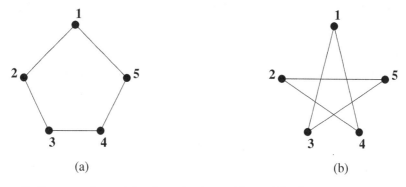

(a)   (b)

In the case of a weighted graph, the problem of finding a minimal-weight Hamiltonian cycle is also known as the *traveling salesman* problem (e.g., Dantzig et al., 1954): Starting from the home city, a salesman has to visit each city (node) on a given list exactly once and then return home. The salesman has to plan the tour such that the total traveling distance (cost) is minimized. A survey of (approximation) algorithms for solving the traveling salesman problem can be found in Jünger et al. (1995). In a cost-sharing context, we return to the traveling salesman problem in chapter 3.

1.1.3 Trees

A connected graph is a *tree* if it contains no cycles. Thus, in a tree a unique path exists between any two nodes. A tree is a classic example of a network that functions as distribution system. A graph with no cycles (i.e., a graph where every component is a tree) is called a *forest*.

A tree that contains every node of the graph is called a *spanning tree*. A spanning tree therefore has $n - 1$ edges. Every connected graph contains a spanning tree. In particular, the complete graph G^N contains n^{n-2} different spanning trees.

A spanning tree represents a way to ensure connectivity between any pair of agents (nodes) without redundant edges. Thus, when edges are costly, the efficient network will be a spanning tree minimizing the total cost.

Let G^N be a complete weighted graph, with weights c_{ij} for every ij. Define a *minimum-weight spanning tree* as a spanning tree minimizing the total weight,

$$T^{min} \in \left\{ \arg \min_{T \in \mathcal{T}_N} \sum_{ij \in T} c_{ij} \right\}, \tag{1.3}$$

where $\mathcal{T}_N$ in (1.3) is the set of n^{n-2} spanning trees associated with G^N. In general, there may be several minimum-weight spanning trees for a given weighted complete graph G^N (for instance, if all weights are identical, all spanning trees have equal total weight). However, if no two weights are the same, T^{min} is unique.

Finding a minimum-weight spanning tree can be done using, for instance, the Prim (1957) or the Kruskal (1956) algorithm sketched below.

Prim: Pick an arbitrary starting node, say, i. Among $N \setminus i$, pick a node j for which c_{ij} is minimal (i.e., $j \in \{\arg \min_{z \in N \setminus i} c_{iz}\}$), and write down the edge ij. Among $N \setminus \{i, j\}$, pick a node l for which c_{il} or c_{jl} is minimal (i.e., $l \in \{\arg \min_{z \in N \setminus \{i,j\}} (c_{iz}, c_{jz})\}$), and add the edge with smallest weight (il or jl); continue this process until the last node has been added.

Kruskal: Order the edges in nondecreasing order according to weight. Pick the first edge in the order, say, ij. Add the next edge in the order if it does not form a cycle with ij; if it forms a cycle with ij, reject it. Go to the next edge in the order, and continue this process until $n - 1$ edges are added.

Note that both algorithms terminate after $n - 1$ steps. Similarly, we can define a *maximum*-weight spanning tree and use the Prim or the Kruskal algorithm with the obvious changes.

Example 1.4 (Finding a minimum-weight spanning tree) Consider the complete weighted graph illustrated below, with $N = \{1, 2, 3, 4\}$, $E = E^N$, and weights given by $(c_{12}, c_{13}, c_{14}, c_{23}, c_{24}, c_{34}) = (1, 2, 4, 1, 3, 5)$.

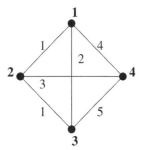

Clearly, there is a unique minimum-weight spanning tree in this case, given by the graph

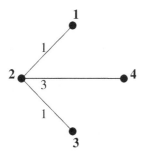

This graph can be found using either:

The Prim algorithm: Start with an arbitrary node, say, node 1. Among the nodes $\{2, 3, 4\}$, clearly $c_{12} = 1$ is the smallest weight of (c_{12}, c_{13}, c_{14}) and hence edge 12 is listed first. Among nodes $\{3, 4\}$, $c_{23} = 1$ is the smallest weight of $(c_{13}, c_{14}, c_{23}, c_{24})$, and hence edge 23 is added to 12. Finally, only node 3 is left, and among (c_{13}, c_{23}, c_{34}), the weight $c_{24} = 3$ is smallest, so edge 24 is added to 12 and 23, and we are done. Thus, the minimum-weight spanning tree consists of nodes $N = \{1, 2, 3, 4\}$ and edges $E = \{12, 23, 24\}$.

or

The Kruskal algorithm: Edges are ordered according to nondecreasing weight as $\{12, 23, 13, 24, 14, 34\}$. Start with edge 12. Then add edge 23, since it does not result in a cycle with 12. Since the next edge 13 results in a cycle with edges 12 and 23, move on to edge 24, which does not result in a cycle with edges 12 and 23, and we are done. As before the minimum-weight spanning tree consists of nodes $N = \{1, 2, 3, 4\}$ and edges $E = \{12, 23, 24\}$.

Consider two $n \times n$ weight matrices K and K'. We say that K' is *smaller than* K (written $K' \leq K$) if and only if $k'_{ij} \leq k_{ij}$ for all $ij \in N$. Now, for a given matrix K and the minimum-weight spanning tree T^{min} of G^N, the *irreducible weight matrix* $K^*(T^{min})$ is defined as the smallest matrix K' such that T^{min} remains the minimum-weight spanning tree. The irreducible weight matrix $K^*(T^{min}) = (k_{ij}^*)$ can be determined as follows. For every $i, j \in N$, let T_{ij}^{min} be the unique path in T^{min} connecting i and j, then $k_{ij}^* = \max_{lm \in T_{ij}^{min}} k_{lm}$.

For a given matrix K, the irreducible weight matrix is unique (even if the complete weighted graph G^N has more than one minimum-weight spanning tree). Consequently, the irreducible weight matrix with respect to T^{min} is simply referred to as K^*.

In principle, the irreducible weight matrix can be defined with respect to an arbitrary spanning tree T in similar fashion. For every $i, j \in N$, let T_{ij} be the unique path in T connecting i and j, then $K^*(T) = (k_{ij}^*)$, where $k_{ij}^* = \max_{lm \in T_{ij}} k_{lm}$.

In line with intuition, we have that $K^*(T^{min}) = K^* \leq K^*(T)$ for any spanning tree T in G^N. (Indeed, we can prove this by contradiction. Assume that there exists $i, j \in N$ for which $\max_{lm \in T_{ij}^{min}} k_{lm} > \max_{lm \in T_{ij}} k_{lm}$, and let lm be such an edge in T_{ij}^{min}. Now, let $T' = T^{min} - lm$. Note that T' is disconnected, and let A_i and A_j be the set of nodes in T' that are connected to i and j, respectively. Then there exists an edge $zh \in E(T_{ij})$ that connects A_i and A_j; let $\widehat{T} = T' + zh$. But then $\widehat{T}$ is a spanning tree for which the total weight is strictly smaller than for T^{min}: a contradiction of our initial assumption.)

Example 1.4 (continued: Finding the irreducible weight matrix) Recall that $T^{min} = \{12, 23, 24\}$ and the weighted adjacency matrix is given by $K = (k_{12}, k_{13}, k_{14}, k_{23}, k_{24}, k_{34}) = (1, 2, 4, 1, 3, 5)$. Thus, we get the irreducible weight matrix $K^* = (k_{12}^*, k_{13}^*, k_{14}^*, k_{23}^*, k_{24}^*, k_{34}^*) = (1, 2, 3, 1, 3, 3)$, since the weights of edges $\{12, 13, 23, 24\}$ remain unchanged, whereas $k_{14}^* = \max\{c_{12}, c_{24}\} = c_{24} = 3$, and $k_{34}^* = \max\{c_{23}, c_{24}\} = c_{24} = 3$.

Now consider, for instance, the spanning tree $T = \{12, 13, 34\}$, with total weight $1 + 2 + 5 = 8$, shown below.

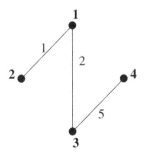

In this case, we get $K^*(T) = (1, 2, 5, 2, 5, 5) \geq (1, 2, 3, 1, 3, 3) = K^*$. Indeed, it can be demonstrated that $K^*(T') \geq K^*$ for any of the remaining $(4^2 - 2 =) 14$ spanning trees T' as well.

In some trees a particular node is called the *root* (or source) and will be denoted 0. Let $G^0 = (N^0, E^0)$ be a *rooted tree*, where $N^0 = N \cup \{0\}$, and $E^0 = E \cup \{i0\}_{i \in N}$. In our models the root is often identical to a supplier in a distribution system: a power plant (or transmission station), a water reservoir, a warehouse, a server in a client-server network, and the like.

Two rooted trees $G^0 = (N^0, E^0)$ and $G^{0'} = (N^{0'}, E^{0'})$ are *isomorphic* if there is a bijection $\lambda : N^0 \to N^{0'}$ such that $ij \in E^0$ if and only if $\lambda(i)\lambda(j) \in E^{0'}$ and $\lambda(0) = 0'$.

1.1.4 Directed Graphs

A graph $G = (N, E)$ is said to be *directed* (a digraph) if each edge has a direction in the sense that it points from one endnode to the other. If the edge $ij \in E$ points from $i \in N$ to $j \in N$, j is said to be the *successor* of i, and i is said to be the *predecessor* of j. A directed edge is often called an *arc*.

The distinction between undirected and directed graphs is often highly important in the models we study. For instance, a given edge may be useful to obtain connectivity in an undirected network, while it can be useless in a directed network if its direction goes against the direction of the network flow. A directed edge may also be used to indicate an order of the endnodes (agents), which in turn may matter for the way resources should be allocated. In terms of network formation, it takes two agents to establish an undirected link (for instance, a trade agreement), while it only takes one agent to establish a directed link (for instance, mailing spam). Clearly, this difference influences the type of network that will arise from agents' interactions.

We can construe a directed graph as being represented by a collection of dominance relations on the set of nodes N. Letting $\mathcal{P}(N)$ denote the set of all subsets of N, a dominance relation is a mapping $D : N \to \mathcal{P}(N)$ such that $i \notin D(i)$ for every $i \in N$. Node $i \in D(j) \subset N$ is *dominating* node j, in the sense that i is an immediate predecessor of j in G. Likewise, $h \in D^{-1}(j)$ is an immediate successor of j in G.

A *directed path* from node i to j in G is a finite sequence of nodes $i_1, \ldots, i_m$ such that $i_1 = i$, $i_m = j$, and $i_k \in D(i_{k+1})$ for $k \in \{1, \ldots, m-1\}$. A directed path for which $i_1 = i_m = i$ is called a *directed cycle*.

Example 1.5 (Hierarchies) A classic *hierarchy* is a cycle-free directed graph for which there exists a unique node i with $D(i) = \emptyset$ (the boss). Consider the graph below.

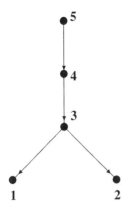

Five agents are organized in a branch hierarchy: agent 5 is the boss; agent 4 is the immediate subordinate; agent 3 is the subordinate of 4; agents 1 and 2 are subordinates of 3 and are the lowest in the hierarchy. In terms of dominance structure, we therefore get: $D(5) = \emptyset$, $D(4) = \{5\}$, $D(3) = \{4\}$, $D(1) = D(2) = \{3\}$.

In a directed graph it is natural to talk about an *in-degree* and an *out-degree* of each node. The in-degree is the number of arcs pointing to the node in question, and the out-degree is the number of arcs pointing from the node in question. In other words, for a given node $i \in N$ the in-degree $\delta^-(i)$ is the number of predecessors of i (i.e., $\delta^-(i) = |D(i)|$). The out-degree $\delta^+(i)$ is the number of successors of i (i.e., $\delta^+(i) = |D^{-1}(i)|$). Clearly, $\sum_{i \in N} \delta^-(i) = \sum_{i \in N} \delta^+(i) = |E|$.

A node $i \in N$ for which $\delta^-(i) = 0$ is called a *source*. A node $i \in N$ for which $\delta^+(i) = 0$ is called a *sink*.

Arcs may also have weights. For instance, suppose each arc in a digraph has a given *capacity* $c_e \geq 0$ representing the maximum flow that can be routed via $e = (i, j)$ in direction from i to j. We can think of flows in connection with traffic, distribution of homogeneous goods (like water, gas, and electricity), information, and so forth. Capacity constraints are important for optimal network design and may also restrict network welfare.

Let s (the source) and t (the sink) be two distinct nodes in N. Then a static *flow* F from s to t is a function $f : E \to \mathbb{R}_+$ satisfying

1. $f(e) \leq c_e$ for all $e \in E$

2. $\sum_{j \in D^{-1}(i)} f(i, j) - \sum_{j \in D(i)} f(j, i) = \begin{cases} F & i = s \\ 0 & i \neq s, t \\ -F & i = t \end{cases}$.

Condition 1 ensures that no flow exceeds the capacity of each arc. Condition 2, often called the conservation property, ensures that the net flow out of any node i is 0 unless i is the source (in which case, the net flow out of s is F) or the sink (in which case the net flow out of t is $-F$).

Finding the maximal flow from source to sink, given the arc capacities of the graph, is called a *max-flow problem* and can be solved using a version of the greedy algorithm as suggested in Ford and Fulkerson (1962):[3]

Step 0. Find an initial $s - t$ directed path, P, in the graph G, and let the flow of this path equal the minimal capacity among arcs in the path (i.e., let $B = \min_{e \in P} c_e$, and assign $f(e) = B$ for all $e \in P$).

Step 1. Construct the residual graph G_f given the flow f as follows. For any arc $e = (i, j)$ with $f(e) > 0$, direct a (backward) arc from j to i with residual capacity $c'_e = f(e)$. Moreover, for any arc $e = (i, j)$ with $f(e) < c_e$, direct a (forward) arc from i to j with residual capacity $c'_e = c_e - f(e)$. Now, find a directed $s - t$ path, P', in the residual graph G_f if such a path exists; otherwise stop.

Step 2. Augment the flow f using P' and minimal capacity $B' = \min_{e \in P'} c'_e$ as follows. Let $f'(e) = f(e) + B'$ if e is an i-to-j directed arc in P', and let $f'(e) = f(e) - B'$ if e is a j-to-i directed arc in P'. Go to step 1 with $f = f'$.

The augmenting loop continues until the residual graph excludes the possibility of an $s - t$ directed flow. Note that using this algorithm, we only find one out of potentially many possible max-flows for a given graph G. For alternative and more recent algorithms, see, for example, Kozen (1992).

A *cut* is a set of arcs $E^* \subseteq E$ separating s and t in G, so $G - E^*$ has max-flow value 0. The *capacity of a cut* is the sum of the capacities for each arc in the cut. By a classic result of Ford and Fulkerson (1962), the capacity of any cut is greater than or equal to the max-flow, and there exists a minimal cut (i.e., a cut with capacity equal to the max-flow).

In the context of revenue sharing, where each arc is controlled by a given agent, we return to max-flow problems in chapter 4.

---

3. Strictly speaking, the current version only works for integer capacities.

Example 1.6 (A max-flow problem) Consider the graph G below, where the numbers indicate the capacity of the respective arcs and arrows indicate the direction of the flow.

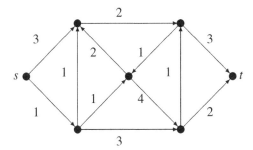

Using the Ford-Fulkerson algorithm to find a max-flow, we proceed as follows.

First, find a $s - t$ directed path in G, for instance, the one resulting in the following flow f (with $B = 1$).

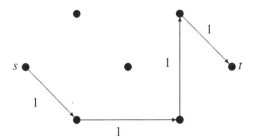

Based on this flow, the residual graph G_f becomes as follows.

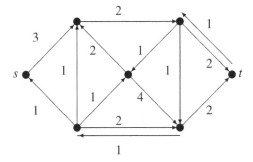

From the residual graph, it is clear that there is an upper $s - t$ directed flow with minimum capacity $B' = 2$, leading to the augmented flow f' below.

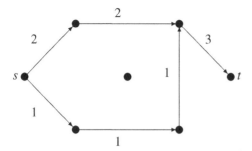

Based on the augmented flow f', we then get residual graph $G_{f'}$ below.

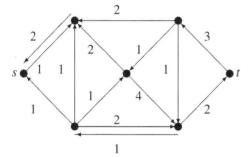

As it appears, there is no $s - t$ directed flow in $G_{f'}$, so the Ford-Fulkerson algorithm terminates, and the flow f' is therefore a max-flow.

It is clear that there is a unique minimal cut, illustrated below using the sign //.

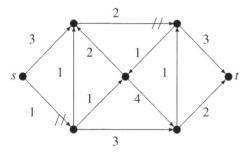

1.2 Cooperative Games

Analyzing allocation problems, economists and game theorists often use the mathematical model of a *cooperative game*. We therefore review a series of basic definitions and results from the theory of *transferable utility* (TU) games (see, e.g., Peleg and Sudhölter, 2007).

1.2.1 Notation and Basic Definitions

A TU game among a set of agents $N = \{1, \ldots, n\}$ models the value (payoff) that any coalition of these agents can obtain in total by participating in the game. It is thus assumed that payoff can be freely reallocated (transferred) among the agents in any coalition. Agents are broadly interpreted as persons, organizations, branches or departments of a corporation, objectives, products, and the like.

A coalition $S \subseteq N$ is a subset of agents. Let $\mathcal{P}(N)$ be the set of all coalitions (subsets) of N. There are $2^n - 1$ such coalitions, disregarding the empty set $\emptyset$.

Let $v : \mathcal{P}(N) \to \mathbb{R}$ be a discrete value function, where $v(S)$ represents the total value (payoff) to coalition $S \in \mathcal{P}(N)$, with $v(\emptyset) = 0$ by definition. Note that a negative value is a *cost*.

Now, a pair (N, v) constitutes a TU game, which we will typically refer to as an *allocation problem* in the present context. Denote by Γ the set of all allocation problems. In our models, cooperation is typically tantamount to being connected in a network. Thus, the game describes what every coalition of agents can obtain by connecting in a network on their own. But notice that a game says nothing about how agents are connected (i.e., the specific configuration of the graph).

Allocation problems are said to be *monotonic* if coalition value is (weakly) increasing in coalition size, that is, if for all $S' \subseteq S \subseteq N$, we have that $v(S') \leq v(S)$.

Two allocation problems (N, v) and (N, w) are said to be *strategically equivalent* if there exists $\alpha > 0$ and $\beta \in \mathbb{R}^N$ such that $w(S) = \alpha v(S) + \sum_{i \in S} \beta_i$ for all $S \subseteq N$

When agents team up, they can coordinate their actions and obtain better resource utilization. One way to model that cooperation is beneficial is to require that the value of joining any two coalitions exceeds the sum of these coalitions' standalone values.

Formally, an allocation problem $(N, v) \in \Gamma$ is said to be *convex* if, for all $S', S \subseteq N$,

$$v(S \cup S') + v(S \cap S') \geq v(S) + v(S'), \tag{1.4}$$

or equivalently, for all i and $S \subseteq T \subseteq N \setminus \{i\}$,

$$v(T \cup i) - v(T) \geq v(S \cup i) - v(S). \tag{1.5}$$

In particular, if condition (1.4) holds for all S, S' where $S \cap S' = \emptyset$, the allocation problem (N, v) is said to be *superadditive*. So convexity implies superadditivity, but superadditivity does not imply convexity.

If, for all S', $S \subseteq N$, it is the case that

$$v(S \cup S') + v(S \cap S') \leq v(S) + v(S'), \tag{1.6}$$

or equivalently, for all i and $S \subseteq T \subseteq N \setminus \{i\}$, that

$$v(T \cup i) - v(T) \leq v(S \cup i) - v(S), \tag{1.7}$$

the allocation problem (N, v) is said to be *concave*. In particular, if condition (1.6) holds for all S, S' where $S \cap S' = \emptyset$, the allocation problem (N, v) is said to be *subadditive*. Since costs are negative values, cooperation is beneficial for any two coalitions in *cost* allocation problems when these are concave (or subadditive).

An *allocation* is a vector $x \in \mathbb{R}^n$ distributing the value $v(N)$ among the n agents, that is, $x_1 + \cdots + x_n = v(N)$. Two equivalent principles for fair allocation are widely studied in the literature (see e.g., Hougaard, 2009).

• *The standalone principle:* Let $x \in \mathbb{R}^n$ be an allocation related to the allocation problem (N, v). Then for every coalition $S \subseteq N$, it is required that

$$\sum_{i \in S} x_i \geq v(S). \tag{1.8}$$

When allocating the social payoff $v(N)$, the standalone principle basically says that no coalition of agents can be given less than what they can obtain on their own. In other words, no coalition of agents has an incentive to block allocations satisfying the standalone principle, since they are better off cooperating and receiving their allocation, $\sum_{i \in S} x_i$, than they are standing alone and receiving their coalitional value $v(S)$. Hence, such allocations can be seen as sustaining cooperation among the agents in N.

Dually, the marginal principle states that no coalition S must be subsidized by agents of the complement $N \setminus S$. Thus, no coalition can be allocated a total value larger than its marginal contribution to the complement $v(N) - v(N \setminus S)$. Formally, we have the following.

• *The marginal principle:* Let $x \in \mathbb{R}^n$ be an allocation related to allocation problem (N, v). Then for every coalition $S \subseteq N$, it is required that

$$\sum_{i \in S} x_i \leq v(N) - v(N \setminus S). \tag{1.9}$$

The marginal principle is equivalent to the standalone principle since

$$\sum_{i \in S} x_i \leq v(N) - v(N \setminus S) = \sum_{i \in N} x_i - v(N \setminus S) \iff \sum_{i \in N \setminus S} x_i \geq v(N \setminus S).$$

This equivalence is convenient, because when agents are, for instance, identified by products (as in some cost allocation problems) and thus cannot actively block cooperation, we can still rely on the fairness notion embodied in the marginal principle.

1.2.1.1 The core

For a given allocation problem (N, v), denote by $core(N, v)$, the set of allocations satisfying the conditions of the standalone principle (or equivalently, those of the marginal principle):

$$core(N, v) = \left\{ x \in \mathbb{R}^n \mid \sum_{i \in N} x_i = v(N), \sum_{i \in S} x_i \geq v(S), \text{ for all } S \subset N \right\}$$

$$= \left\{ x \in \mathbb{R}^n \mid \sum_{i \in N} x_i = v(N), \sum_{i \in S} x_i \leq v(N) - v(N \setminus S), \right. \quad (1.10)$$

$$\left. \text{for all } S \subset N \right\}.$$

Mathematically, the core is a compact and convex subset of $\mathbb{R}^n$. The conditions (1.8) or (1.9) will often be referred to as the *standalone core conditions* or just the *core conditions*.

Obviously, there may be situations where some coalitions violate the standalone principle, so the core may be empty. If the core is empty, it means that there is no possibility of sharing the total value that sustains a joint action of all agents (e.g., if there are no advantages from cooperation).

Looking for necessary and sufficient conditions for a nonempty core, we can consider the following linear program:

$$\min \sum_{i \in N} x_i \text{ s.t. } \sum_{i \in S} x_i \geq v(S), \quad \text{for all } S \in \mathcal{P}(N). \quad (1.11)$$

Clearly, $core(N, v) \neq \emptyset$ if and only if the optimal solution to (1.11) is equal to $v(N)$, with any such solution being a core allocation. By linear programming duality, we can obtain an equivalent necessary and sufficient condition for a nonempty core based on *balancedness*: Given the set of agents N, a collection $\mathcal{B} = \{S_1, \ldots, S_m\}$ of nonempty subsets of N is called *balanced* if there exist positive scalars, $\delta_1, \ldots, \delta_m$, such that

$$\sum_{j:i \in S_j} \delta_j = 1, \quad \text{for every } i \in N.$$

The collection δ is called a system of *balancing weights*.

Every partition of N is a balanced collection; consequently, balanced collections can be viewed as generalized partitions. The total value of all such balanced arrangements plays a crucial role in relation to the existence of core elements, since, by the Bondareva/Shapley Theorem, we have that $core(N, c) \neq \emptyset$ *if and only if, for all systems of balancing weights δ,*

$$\sum_{S \subseteq \mathcal{B}} \delta_S v(S) \le v(N). \tag{1.12}$$

Consequently, if an allocation problem (N, v) has nonempty core, it is said to be balanced. Now, denote by v^S the $|S|$−agent problem obtained by restricting v to coalitions $T \subseteq S$. A problem (N, v) is said to be *totally balanced* if and only if all its subproblems (S, v^S), $S \in \mathcal{P}(N)$, are balanced. For instance, convex problems are totally balanced.

Example 1.7 (Three-agent problems with nonempty core) In the case of three-agent problems, $N = \{1, 2, 3\}$, all partitions of $\{1, 2, 3\}$ are balanced as well as the collection $(\{1, 2\}, \{1, 3\}, \{2, 3\})$. Balancing weights are given as solutions to the equations

$\delta_1 + \delta_{12} + \delta_{13} = 1,$

$\delta_2 + \delta_{12} + \delta_{23} = 1,$

$\delta_3 + \delta_{13} + \delta_{23} = 1.$

Hence, the core is nonempty for three-agent allocation problems if and only if

$v(1) + v(2, 3) \le v(1, 2, 3),$

$v(2) + v(1, 3) \le v(1, 2, 3),$

$v(3) + v(1, 2) \le v(1, 2, 3),$

$v(1) + v(2) + v(3) \le v(1, 2, 3),$

$0.5(v(1, 2) + v(1, 3) + v(2, 3)) \le v(1, 2, 3).$

It appears that superadditivity of (N, v) is not enough to guarantee a nonempty core (it only implies that the first four conditions are satisfied). The last condition is satisfied if (N, v) is convex (in fact, convexity turns out to be a sufficient condition for a nonempty core in general).

The core is not the only relevant set of allocations, even though it is well founded on the standalone principle. Supersets of the core can also be relevant, for example, the so-called *Weber set.*

To define the Weber set, let $\pi : N \to N$ be an ordering of the agents in N, and let Π be the set of all such $n!$ orderings. Let $S_{\pi,i} = \{j \in N \mid \pi(j) \leq i\}$ be the coalition of agents up to (and including) agent i in the ordering π. For each ordering $\pi \in \Pi$, and agent $i \in N$, define the marginal contribution of i, given π, as

$$x_i^\pi = v(S_{\pi,i}) - v(S_{\pi,i-1}). \tag{1.13}$$

The Weber set of allocation problem (N, v) is defined as

$$web(N, v) = conv\{x_i^\pi \mid i \in N, \pi \in \Pi\}, \tag{1.14}$$

where $conv(\cdot)$ is the convex hull operator. The Weber set is a *core cover*, that is, $core(N, v) \subseteq web(N, v)$. Furthermore, an allocation problem (N, v) is convex if and only if $core(N, v) = web(N, v)$.

Network design problems arguably fit the TU game model well, and the core becomes a central benchmark for fair allocation in this case. We will see several examples of that in the coming chapters, primarily in chapters 2 and 3.

1.2.2 Allocation Rules

If the gains from cooperation are large enough, there are many ways to share the common value satisfying the standalone principle. The challenge becomes how to select specific allocations.

An *allocation rule* is a function ϕ that assigns to every allocation problem (N, v) a unique payoff vector $\phi(N, v) \in \mathbb{R}^n$ satisfying *budget-balance*:

$$\sum_{i \in N} \phi_i(N, v) = v(N). \tag{1.15}$$

Here we consider two classic examples of an allocation rule: the *nucleolus* and the *Shapley value*. The nucleolus is an example of a rule selecting a core allocation (when the core is nonempty), while the Shapley value is not guaranteed to satisfy the standalone principle.

1.2.2.1 The nucleolus

Consider a given allocation problem (N, v), and let x be a specific allocation (i.e., $x_1 + \cdots + x_n = v(N)$). Given (N, v) and x, for every coalition $S \subset N$, let

$$e_S(x) = \sum_{i \in S} x_i - v(S) \tag{1.16}$$

be the gains of coalition S from a cooperative action among the agents in N. Note that $e_S(x) \geq 0$, for all S, if and only if x satisfies the standalone principle.

Now, given the allocation x, coalition S is said to be better off than coalition T if S gains more than T from cooperation among N (i.e., if $e_S(x) > e_T(x)$).

Let $e(x) \in \mathbb{R}^{2^n-2}$ be the vector of gains given the allocation x (disregarding coalitions N and $\emptyset$), and let $\theta : \mathbb{R}^{2^n-2} \to \mathbb{R}^{2^n-2}$ map vector elements in increasing order. The nucleolus of allocation problem (N, v) is defined as the allocation that lexicographically maximizes $\theta(e(x))$. That is, the nucleolus is defined by the allocation rule

$$\phi^{Nuc}(N, v) = \{x \in I(N, c) \mid \theta(e(x)) \succ_{lex} \theta(e(y)), \ \forall y \in I(N, c)\}, \tag{1.17}$$

where $I(N, c) = \{x \in \mathbb{R}^n \mid \sum_{i \in N} x_i = v(N)\}$, and $\succ_{lex}$ is the lexicographic ordering given by $x \succ_{lex} y$ if $x_1 > y_1$, or if $x_1 = y_1$ and $x_2 > y_2$, and so on. In other words, the nucleolus is the allocation that favors the worst-off coalition in terms of gains from central coordination.

For any *balanced* allocation problem (N, v), it can be shown that $\phi^{Nuc}(N, v) \in \text{core}(N, v)$. Consequently, ϕ^{Nuc} is a core selection.

In general, however, there are also good arguments in favor of selections that potentially may be in conflict with the standalone principle. An example of such a selection is the Shapley value.

1.2.2.2 The Shapley value

For each agent i, let $m_i(S) = v(S \cup i) - v(S)$ be the marginal contribution of i to coalition S. Intuitively, it seems reasonable that agents should be rewarded according to their contribution to the social welfare. But the (marginal) contribution of a given agent depends on which group she joins. Thus, one way to allocate $v(N)$, using these marginal contributions $m_i(S)$, is to take a weighted average of m_i over all possible coalitions $S \subseteq N \setminus i$ that i can join. Then define the *Shapley value*, ϕ^{Sh}, for allocation problem (N, v) as

$$\phi_i^{Sh}(N, v) = \sum_{S \subseteq N \setminus i} \frac{s!(n-s-1)!}{n!} m_i(S), \quad \text{for all} \quad i \in N, \tag{1.18}$$

(where $|S| = s$, and $0! = 1$). The Shapley value is the barycenter of the Weber set, since $\phi_i^{Sh}(N, v) = \frac{1}{n!} \sum_{\pi \in \Pi} x_i^\pi$, for all $i \in N$. Consequently, convexity of (N, v) is a sufficient condition for $\phi^{sh}(N, v) \in core(N, v)$. Moreover, the Shapley value and the nucleolus coincide for two-agent problems.

Example 1.8 (Two problems with the same core but different Nucleoli and Shapley values) Let $N = \{1, 2, 3, 4\}$, and let $v(S)$ for $S \subseteq N$ be defined by

$$v(N) = 2,$$

$$v(\{1, 2, 3\}) = v(\{1, 2, 4\}) = v(\{1, 3, 4\}) = v(\{2, 3, 4\}) = 1,$$

$v(\{1,2\}) = v(\{3,4\}) = v(\{1,4\}) = v(\{2,3\}) = 1,$

$v(\{1,3\}) = 0.5, \ v(\{2,4\}) = 0,$

$v(\{i\}) = 0 \ \text{for} \ i = 1, \ldots, 4.$

Additionally, define the allocation problem (N, v^*) by $v^*(S) = v(S)$ for all $S \neq \{1, 2, 3\}$ and $v^*(\{1, 2, 3\}) = 1.25$.

First, note that both problems (N, v) and (N, v^*) are balanced but not convex. Further, both problems have the same core:

$$core(N, v) = core(N, v^*) = conv\{(0.25, 0.75, 0.25, 0.75), (1, 0, 1, 0)\}.$$

Then

$$\phi^{Nuc}(N, v) = (0.5, 0.5, 0.5, 0.5),$$

$$\phi^{Sh}(N, v) = (0.54, 0.46, 0.54, 0.46).$$

So the nucleolus and the Shapley value differ on (N, v), but both allocations satisfy the standalone principle in this case.

Further, we get that

$$\phi^{Nuc}(N, v^*) = (0.625, 0.375, 0.625, 0.375) \neq (0.5, 0.5, 0.5, 0.5)$$

$$= \phi^{Nuc}(N, v),$$

$$\phi^{Sh}(N, v^*) = (0.56, 0.48, 0.56, 0.40) \neq (0.54, 0.46, 0.54, 0.46) = \phi^{Sh}(N, v).$$

So in both cases the allocations differ between the problems, even though the core is the same. Moreover, the Shapley value does not satisfy the standalone principle in the latter case since we have that $\phi_1^{Sh}(N, v^*) + \phi_4^{Sh}(N, v^*) = 0.96 < 1 = v(1, 4)$.

The literature offers several axiomatic characterizations of both the nucleolus and the Shapley value. Although they have a common foundation for two-agent problems, consistency with respect to different types of reduced games leads to radically different solutions in general. Another difference relates to the property of additivity (here stating that $\phi(N, v' + v') = \phi(N, v) + \phi(N, v')$): the Shapley value is additive; the nucleolus is not. A third difference relates to monotonicity properties as shown in the next section.

1.2.3 Monotonicity versus the Standalone Principle

To justify the possibility of relevant properties of allocation rules that fail to comply with the standalone principle, consider the following monotonicity requirement.

If, for all agents $i \in N$, the value increases for all coalitions containing agent i (the value for all other coalitions being kept fixed), then i should not be allocated less. Formally, an allocation rule ϕ is *coalitionally monotonic* if, for all $i \in N$,

$$[v(S) \geq \hat{v}(S) \text{ for all } S \text{ containing } i, \text{ and } v(T) = \hat{v}(T) \text{ otherwise}]$$
$$\Rightarrow \phi_i(N, v) \geq \phi_i(N, \hat{v}). \tag{1.19}$$

Alternatively, if the value increases for a given coalition (the values of all other coalitions being kept fixed), then the resulting allocation of any member of that coalition should not decrease. Formally, for a particular coalition $S \subset N$, an allocation rule ϕ is *S-monotonic* if, for all $i \in S$,

$$[v(S) \geq \hat{v}(S), \text{ and } v(T) = \hat{v}(T) \text{ for all } T \neq S]$$
$$\Rightarrow \phi_i(N, v) \geq \phi_i(N, \hat{v}). \tag{1.20}$$

The allocation rule ϕ is coalitionally monotonic if and only if ϕ is S-monotonic for all coalitions $S \subseteq N$.

Broadly speaking, the standalone principle can be viewed as ensuring that cooperation is not blocked by conflicting interests, as *given* by the participation constraints of all coalitions, while the property of coalitional monotonicity relates to conflicting interests among members of a coalition when the participation constraints *vary*. As it turns out, there is a trade-off between these conditions, in the sense that, no allocation rule can satisfy both the standalone principle and coalitional monotonicity on the class of balanced allocation problems. *For $n \geq 4$, no core allocation rule satisfies coalitional monotonicity on the class of balanced problems.*

Consequently, the nucleolus is not coalitionally monotonic on the class of balanced allocation problems. In fact, the nucleolus is not even N-monotonic. In contrast, the Shapley value is clearly coalitionally monotonic by the way it is defined, but it is not a core selection on the class of balanced problems. Note, however, that on the subclass of convex allocation problems, the Shapley value is an example of a core selection that is coalitionally monotonic.[4]

Since coalitional monotonicity is a desirable property based on considerations of fairness, as well as of incentives, there can be good arguments in favor of allocations that do not necessarily comply with the standalone principle.

---

4. See also Hougaard and Østerdal (2010) for further analysis of S-monotonicity in the case where a social planner is imagined to maximize a (strictly concave and differentiable) welfare function over the set of core allocations.

1.3   Graphs and Games

Let us now briefly consider some interconnections between graphs and games. In particular, we discuss how an initial game can be altered (or restricted) by a graph describing permissible cooperation structures, and conversely, how allocation problems related to an initial graph (network structure) can be modeled as cooperative games.

1.3.1   Games with Graph Restrictions

The seminal approach to graph restrictions is due to Myerson (1977), who suggested that graphs can be used to represent partial cooperation structures in TU games. Assume that cooperation takes place as a series of bilateral agreements among the agents. When two agents agree, a connection is formed, and the set of all permissible cooperation structures can therefore be identified by the set of all possible graphs connecting agreeing agents.

Fix N, and let $\mathcal{G} = \{G \mid G \subseteq G^N\}$ be the set of all possible graphs. Given a graph $G \in \mathcal{G}$ and a coalition $S \subseteq N$,

$$S^G = \{\{i \mid i \text{ and } j \text{ are connected in } S \text{ by } G\} \mid j \in S\} \tag{1.21}$$

defines the unique partition of S into groups that are connected in S given the structure of the graph G. For every coalition $S \subseteq N$, the partitions S^G can be used to define a natural graph restriction on a game v.

Consider a given initial game v and a graph G, representing a cooperation structure. Let, for all coalitions $S \subseteq N$,

$$v^G(S) = \sum_{T \in S^G} v(T) \tag{1.22}$$

define a new game where agents are restricted to cooperate as represented by the graph G. That is, the value of each coalition S is simply the sum of the standalone values for each subgroup in the associated partition S^G. The restricted game v^G is balanced (i.e., has nonempty core) if v is superadditive and the graph G contains no cycles.

Loosely speaking, this observation seems to justify communication structures in the form of hierarchies (trees), since these structures enhance the chances of reaching stable allocations in the sense of the core. Given a hierarchy, the underlying game v only needs to be superadditive to guarantee a nonempty core of the restricted game (1.22).[5]

---

5. See, e.g., Demange (1994, 2004) for further analysis and discussion on this issue.

Example 1.9 (A graph restricted game) Recall the four-agent game v defined in example 1.8, and consider the (disconnected) graph G below.

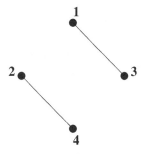

Using definition (1.22) results in the following graph restricted game:

$$v^G(1, 2, 3, 4) = v(1, 3) + v(2, 4) = 0.5,$$
$$v^G(1, 2, 3) = v(1, 3) + v(2) = 0.5,$$
$$v^G(1, 2, 4) = v(2, 4) + v(1) = 0,$$
$$v^G(1, 3, 4) = v(1, 3) + v(4) = 0.5,$$
$$v^G(2, 3, 4) = v(2, 4) + v(3) = 0,$$
$$v^G(1, 2) = v(1) + v(2) = 0,$$
$$v^G(1, 3) = v(1, 3) = 0.5,$$
$$v^G(1, 4) = v(1) + v(4) = 0,$$
$$v^G(2, 3) = v(2) + v(3) = 0,$$
$$v^G(2, 4) = v(2, 4) = 0,$$
$$v^G(3, 4) = v(3) + v(4) = 0,$$
$$v^G(i) = v(i) = 0, \quad \text{for all } i.$$

Clearly the restricted game v^G is very different from the original game v. For instance, $v^G(1, 2, 3, 4) = 0.5 < 2 = v(1, 2, 3, 4)$. Note that $v^G(N) = v(N)$ if and only if G is connected. Note further that since the coalition consisting of agents 2 and 4 cannot generate value by standing alone in the original game (i.e., $v(2, 4) = 0$), and the graph G isolates them as a component, the stand-alone conditions of the restricted game v^G do not allow these agents to get any payment contrary to the core of the original game.

Myerson (1977) calls an allocation rule *fair* if the absolute change in payoff from removing (adding) an edge between agents i and j is the same for both agents. In other words, two agents should gain equally from entering a bilateral agreement (connection). Formally, ϕ is fair if for all graphs $G \in \mathcal{G}$, and for all edges $ij \in G$, we have

$$\phi_i(N, v^G) - \phi_i(N, v^{G-ij}) = \phi_j(N, v^G) - \phi_j(N, v^{G-ij}). \tag{1.23}$$

The Shapley value of the graph restricted game, $\phi^{Sh}(N, v^G)$, is the unique fair allocation rule, often dubbed the *Myerson value*.

Although superadditivity of the game v together with a cycle-free graph G creates a balanced graph restricted game, there is no guarantee that the Myerson value satisfies the core conditions (since the graph restricted game need not be convex).

For graphs without cycles, Herings et al. (2008) therefore suggest weakening the fairness condition to so-called *component fairness*:

$$
\frac{1}{|K^i|} \sum_{h \in K^i} (\phi_h(N, v^G) - \phi_h(N, v^{G-ij}))
$$

$$
= \frac{1}{|K^j|} \sum_{h \in K^j} (\phi_j(N, v^G) - \phi_j(N, v^{G-ij})), \tag{1.24}
$$

where K^i is the component of $G - ij$ containing node i, and K^j is the component of $G - ij$ containing node j. That is, component fairness states that removing the edge between agents i and j on average gives the agents in the resulting two components, K^i and K^j, the same payoff.

On the class of cycle-free graph restricted games, there exists a unique component fair allocation rule, dubbed the *average tree (AT) solution*.

The AT solution has a nice intuitive interpretation. Consider a given tree T (or component if G is a forest), and the let each node (agent) in turn take the role as root of the tree. This creates n hierarchies, T^i, where the root i is the unique "boss" (i.e., $D(i) = \emptyset$), and the rest of the tree indicates a structure of subordinates. For each such hierarchy, we can define a marginal payoff vector for agent j when i is the root:

$$
m(j:i) = v(S_{T^i}(j)) - \sum_{h \in D^{-1}(j)} v(S_{T^i}(h)),
$$

where $S_{T^i}(j)$ is the coalition of all subordinated agents of the *sub*tree of T^i where j is the root (including j itself), and $D^{-1}(j)$ is the set of immediate subordinates of j in T^i (i.e., those subordinated agents of j that is directly linked to j in T^i). The AT payoff to any agent j is the average of these marginal payoffs $m(j:i)$ over all hierarchies:

$$AT_j(v^G) = \frac{1}{|K|} \sum_{i \in K} m(j:i), \tag{1.25}$$

where j belongs to component K of the forest G.

The AT solution satisfies the core conditions of the graph restricted game, that is, $AT(v^G) \in core(v^G)$.

Example 1.10 (Computing the Myerson value and the AT solution) Reconsider the game v from example 1.8. This time let the graph G be given as the tree (component) $G = \{12, 23\}$ resulting in the graph restricted game:

$$v^G(1, 2, 3, 4) = v(1, 2, 3) + v(4) = 1,$$

$$v^G(1, 2, 3) = v(1, 2, 3) = 1,$$

$$v^G(1, 2, 4) = v(1, 2) + v(4) = 1,$$

$$v^G(1, 3, 4) = v(1, 3) + v(4) = 0,$$

$$v^G(2, 3, 4) = v(2, 3) + v(4) = 1,$$

$$v^G(1, 2) = v(1, 2) = 1,$$

$$v^G(1, 3) = v(1) + v(3) = 0,$$

$$v^G(1, 4) = v(1) + v(4) = 0,$$

$$v^G(2, 3) = v(2, 3) = 1,$$

$$v^G(2, 4) = v(2) + v(4) = 0,$$

$$v^G(3, 4) = v(3) + v(4) = 0,$$

$$v^G(i) = v(i) = 0, \quad \text{for all } i.$$

Clearly, $core(N, v^G) = (0, 1, 0, 0)$. The Myerson value is given by

$$\phi^{Sh}(v^G) = \left(\frac{1}{6}, \frac{2}{3}, \frac{1}{6}, 0\right),$$

which differs from the (unique) core allocation.

Now, consider the AT solution. First we find the marginal payoff vectors of the tree component $T = \{12, 23\}$ for each of the following three hierarchies.

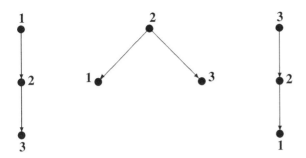

These payoffs are

$$m(1:1) = v^G(1, 2, 3) - v^G(2, 3) = 0,$$

$$m(2:1) = v^G(2, 3) - v^G(3) = 1,$$

$$m(3:1) = v^G(3) - v^G(\emptyset) = 0,$$

$$m(1:2) = v^G(1) - v^G(\emptyset) = 0,$$

$$m(2:2) = v^G(1, 2, 3) - v^G(1) - v^G(3) = 1,$$

$$m(3:2) = v^G(3) - v^G(\emptyset) = 0,$$

$$m(1:3) = v^G(1) - v^G(\emptyset) = 0,$$

$$m(2:3) = v^G(1, 2) - v^G(1) = 1,$$

$$m(3:3) = v^G(1, 2, 3) - v^G(1, 2) = 0.$$

Thus, for the component T, we get the AT solution

$$AT_i = \frac{1}{3}(m(i:1) + m(i:2) + m(i:3))$$

for $i = 1, 2, 3$.

Since the remaining component, consisting of the singleton node $\{4\}$, has value $v^G(4) = 0$, we have the AT solution of the graph restricted game (N, v^G)

$$AT(N, v^G) = (0, 1, 0, 0),$$

coinciding with the unique core allocation.

The literature has several variations over the Myerson approach, for instance, so-called *games with permission structure* (see, e.g., Gilles et al. 1992 and the book Gilles 2010).

1.3.2 From Graphs to Games

In Myerson's approach above, the value of cooperation is fixed by the initial TU game (which is exogenously given), and the role of the graph resulting from pairwise agreements is only to restrict this (already fixed) value by restricting the structure of potential coalitions. However, in many cases it seems more relevant to turn things upside down and let networks be the primitives of the model, such that value is the direct result of network structures rather than of coalitions of agents. This is a much richer framework, since, for instance, the same coalition of agents can be connected in different network structures, and these may have different values for the coalition. In other words, the value of a given coalition in a network, or network design problem, is not uniquely defined but depends on multiple factors, such as the specific network configuration, the network configuration of agents not in the coalition, and whether agents in the coalition can connect via agents outside the coalition.

In the coming chapters we will examine several models that attempt to represent a given network allocation problem as a cooperative game and demonstrate how different definitions of the game lead to different allocations with different characteristics, even using the same type of allocation rule. So the main challenge is how to go from a given network allocation problem to a specific TU game, and it becomes crucial to evaluate whether this transformation captures the type of allocation issues we aim to address.

The relevance of the game representation typically depends on whether we consider allocation problems in a given existing network, or in a network design problem, where allocation takes place after determination of the optimal network.

In the latter case, a substantial body of literature documents the usefulness of the game-theoretic approach under complete information, where agents are prevented from acting strategically.

However, in the former case, the literature is much more sparse. In existing networks the standalone costs of the various coalitions are typically counterfactual, since forming a subnetwork serving coalitional connection demands is not an option, at least in the short run. Thus, it is questionable whether the arguments behind the standalone core conditions have any relevance in this case. Furthermore, even if we do accept fairness in the form of the core conditions, existing networks may also contain redundant connections, for which the agents nevertheless are liable, and the presence of such redundant connections typically causes the core of the derived game to be empty. In the coming chapters we will see several examples of network allocation problems that do not have a natural representation as a cooperative game. In such cases, fairness notions and relevant allocation rules have to be directly related to the specific network structure as well as to agent preferences (to the extent these are known).

Suppose it is reasonable to represent a given network allocation problem as a cooperative game. The next challenge is then to define the specific coalitional values: a game models every coalition's worth when standing alone, but typically there are multiple interpretations of what such standalone values are for a given network. To see this, consider the following example involving a simple chain-structured graph.

Example 1.11 (Multiple game interpretations)   Imagine a closed side street in winter, where the houses along the street are responsible for keeping the sidewalks free from snow. Every house owner has to clean their part of the sidewalk in front of the house. Assume there are three houses on the street starting with house number 1 (closest to the main road 0) and ending with house number 3, modeled by the following chain.

The length of each edge corresponds to the length of the sidewalk in front a house, with the edge 01 representing the length of the sidewalk in front of house number 1, the edge 12 representing the length of the sidewalk in front of house number 2, and so forth.

Each edge has a weight expressed as the cost (in terms of labor time or disutility) of the relevant house owner from shoveling snow. Suppose that $k_i > 0$ is the cost connected with the sidewalk of house owner i.

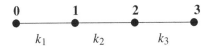

Say that Ann, Bob, and Carl are the respective house owners. There are externalities from cleaning sidewalks, since both Bob and Carl benefit from Ann's clean sidewalk when walking toward the bus stop at main road, but note that Ann does not benefit from any of their shoveling in this case. So we can claim that externalities are asymmetric in the sense that they only exist for agents down the road.

If externalities are present, it can be argued that it makes sense to reallocate the total costs between house owners. For example, Bob may be willing to compensate Ann for part of her disutility, since Ann's job actually benefits

Bob, while Ann is indifferent between Bob cleaning his sidewalk or not and consequently is unwilling to compensate Bob.

Consequently, considered as a game with N agents, there are at least two different (polar) ways to model the situation represented by the chain structured network. First, it can be argued that any coalition of house owners is responsible for cleaning their own sidewalks, and hence the total cost of a coalition equals the sum of the individual costs. That is, for all $S \subseteq N$,

$$c(S) = \sum_{i \in S} k_i. \tag{1.26}$$

Second, focusing on the asymmetric externality involved, it can be argued that any coalition is responsible for cleaning the sidewalk all the way from the main road to the member farthest away from the main road (the agent with the highest index in S: max S). That is, for all $S \subseteq N$,

$$\hat{c}(S) = \sum_{i=1}^{\max S} k_i. \tag{1.27}$$

Obviously it matters for the allocation of costs, which of the two games we consider. Using, for example, the Shapley value, ϕ^{Sh}, we get

$$\phi_i^{Sh}(N, c) = k_i, \quad \text{for all} \quad i \in N,$$

and

$$\phi_i^{Sh}(N, \hat{c}) = \sum_{j=1}^{i} \frac{k_j}{n - j + 1}, \quad \text{for all} \quad i \in N.$$

So in the latter case, the costs of each incremental part are shared equally in a serial fashion, satisfying fairness with respect to the externality of agents down the road, while in the former case the Shapley value makes each house owner cover their own (incremental) cost.

We will return to the case of chain-structured graphs in chapter 2 (although in slightly different versions). For example, we will look at the so-called *airport problem*, where runway costs are allocated among aircraft types on the basis of the fact that larger aircraft require longer runways. Therefore larger aircraft have a positive (cost) externality from the segment of the runway already designed for smaller aircraft, which should be taken into account when allocating costs, as in game (1.27) in example 1.11. In contrast, the game defined in (1.26) does not make sense for the airport problem, as will become clear.

In general network structures (e.g., imagine complete graphs), multiplicity of game interpretations may also arise from whether we allocate among nodes or edges (as in the traveling salesman and Chinese postman games mentioned in section 1.1) or whether there are enforceable ownership rights. Imagine, for example, that a group of agents wants connection to a common supplier (the source) and that these agents are going to share the cost of designing an efficient (cost-minimizing) network connecting everybody to the source (directly or indirectly via other agents). When finding the coalitional value for a given coalition, S, it matters whether agents in S are allowed to connect to the source via other agents outside S or not (i.e., whether agents outside S can enforce their property rights and prevent members of S from connecting to the source through their node). We return to such issues several times in this book, for instance, when analyzing the minimum-cost spanning tree model (as sketched above) in chapter 2.

1.4 Exercises

1.1. Find the degree of centrality (as defined in (1.2)) for each node in the graph below.

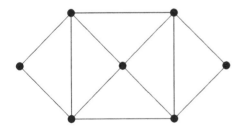

Is the degree of centrality a good measure for how well connected agents (here nodes) are in a network? Suggest some alternative measures.

1.2. Consider the same graph as in exercise 1.1. Use Fleury's algorithm to find an Euler cycle. Is your Euler cycle a circuit? Are there subgraphs for which Euler circuits exist? Find Hamiltonian cycles. Try to formulate an algorithm for finding a Hamiltonian cycle in this case.

1.3. A *bipartite graph* is a graph $G = (N, E)$, where N can be partitioned into two sets N_1 and N_2 ($N_1 \cup N_2 = N$ and $N_1 \cap N_2 = \emptyset$) such that every edge joins a node from N_1 with a node from N_2. Prove that G is bipartite if and only if G does not have an odd length cycle.

1.4. Let $N = \{1, 2, 3\}$, and consider the rooted complete graph G^{N^0}. Determine the number of nonisomorphic spanning tree configurations. Also determine the number of isomorphic graphs for each such type of configuration. Now add an agent (such that $|N^0| = 5$); then there are nine nonisomorphic spanning trees: Illustrate them. How many chains are there (in general)?

1.5. Construct a complete weighted graph with root 0, and suppose that the weights are given by the costs associated with maintaining edges in the graph. Use both the Prim and the Kruskal algorithms to suggest ways of allocating the total cost of a minimum-cost spanning tree when N identifies the set of agents. What would be the result of applying these suggestions to the case presented in example 1.4 (with node 1 as root)?

1.6. Prove that the irreducible weight matrix K^* is unique for a given weight matrix K.

1.7. Continue exercise 1.5 (building on example 1.4) by representing the allocation problem as a TU game (note that it is a cost game). Is the game concave? Find the core. Find the nucleolus and the Shapley value. Compare with your suggested solutions from exercise 1.5.

1.8. Prove that if v is superadditive, then the Shapley value satisfies individual rationality, that is, $\phi_i^{Sh}(N, v) \geq v(\{i\})$ for all $i \in N$. Moreover, prove that $\phi^{Nuc} \in core(N, v)$ when $core(N, v) \neq \emptyset$.

1.9. Consider the game defined in example 1.8 and the graph $G = \{12, 23, 24\}$. Compute the Myerson value and the AT solution. Compare the results.

1.10. Consider a simple linear hierarchy (digraph) where a boss has a subordinate, who herself has a subordinate, and so forth. Each agent i generates a revenue r_i, and the total revenue of the hierarchy $\sum_i r_i$ has to be allocated among the agents in the hierarchy. Discuss whether this allocation problem can be modeled as a cooperative game. If yes, how should the game be defined?

1.11. Consider the max-flow problem of example 1.6. Find all possible paths with a strictly positive flow. Which of those paths result in the max-flow? How can we change the direction of the arcs (keeping capacities fixed) such that there are multiple minimal cuts in this particular max-flow problem? Discuss how to measure the importance of specific arcs in obtaining a given max-flow.

References

Bollobas, B. (1979), *Graph Theory: An Introductory Course*. New York: Springer.

Bollobas, B. (2002), *Modern Graph Theory*. New York: Springer.

Dantzig, G. B., D. R. Fulkerson, and S. M. Johnson (1954), "Solution of a Large Scale Traveling-Salesman Problem" *Operations Research* 2: 393–410.

Demange, G. (1994), "Intermediate Preferences and Stable Coalition Strutures," *Journal of Mathematical Economics* 23: 45–58.

Demange, G. (2004), "On Group Stability in Hierarchies and Networks," *Journal of Political Economy* 112: 754–778.

Eiselt, H. A., M. Gendreau, and G. Laporte (1995), "Arc Routing Problems, Part I: The Chinese Postman Problem," *Operations Research* 43: 231–242.

Ford, L. R., and D. R. Fulkerson (1962), *Flows in Networks*. Princeton, NJ: Princeton University Press.

Gilles, R. P. (2010), *The Cooperative Game Theory of Networks and Hierarchies*. New York: Springer.

Gilles, R. P., G. Owen, and R. van den Brink (1992), "Games with Permission Structures: The Conjunctive Approach," *International Journal of Game Theory* 20: 277–293.

Herings, J. J., G. van der Laan, and D. Talman (2008), "The Average Tree Solution for Cycle-Free Graph Games," *Games and Economic Behavior* 62: 77–92.

Hougaard, J. L. (2009), *An Introduction to Allocation Rules*. New York: Springer.

Hougaard, J. L., and L. P. Østerdal (2010), "Monotonicity of Social Welfare Optima," *Games and Economic Behavior* 70: 392–402.

Jünger, M., G. Reinelt, and G. Rinaldi (1995), "The Traveling Salesman Problem," In M. O. Ball et al., (eds) *Handbooks in OR & MS, vol 7. Handbooks in Operations Research and Management Science*, vol. 7. Amsterdam: North-Holland.

Kozen, D. C. (1992), *The Design and Analysis of Algorithms*. New York: Springer.

Kruskal, J. B. (1956), "On the Shortest Spanning Subtree of a Graph and the Traveling Salesman Problem," *Proceedings of the American Mathematical Society* 7: 48–50.

Kwan, M. K. (1962), "Graphic Programming Using Odd or Even Points," *Chinese Mathematics* 1: 273–277.

Myerson, R. B. (1977), "Graphs and Cooperation in Games," *Mathematics of Operations Research* 2: 225–229.

Peleg, B., and P. Sudhölter (2007), *Introduction to the Theory of Cooperative Games*, 2nd ed. Dordrecht: Kluwer.

Prim, R. C. (1957), "Shortest Connection Networks and Some Generalizations," *Bell System Technical Journal* 36: 1389–1401.

2 Trees

Networks in the shape of trees are often the result of efficient network design. Indeed, if a connected graph contains a cycle, it remains connected when deleting any single edge in the cycle. Hence, if edges can be used without rivalry and are costly, tree structures tend to emerge.

The efficiency of the tree as a distribution system is witnessed by numerous examples from our daily lives. For instance, tree structures appear in irrigation ditches and other water distribution systems, as well as in radial electrical transmission systems. Trees have also proved to be efficient in distributing information, for example, in computer and sensor networks, or in hierarchical organization structures with chains of command.

A tree has a unique path connecting every pair of nodes. In particular, this property implies that in a rooted tree, every node has a unique path to the root (see chapter 1). This fact makes it relatively simple to determine individual agent's network usage when the root acts as supplier, and it plays a key role in many of the allocation rules we will encounter. The associated notions of fairness are often directly related to this fact.

However, we will also see that the situation becomes much more complex with seemingly minor modifications of the model, allowing, for instance, for public (Steiner) nodes or various forms of externalities (e.g., congestion).

Overall, we study two main types of allocation problems:

• Allocation problems associated with an existing network structure which, at least in the short run, must be taken as fixed;

• Allocation problems associated with a network design problem, where information about alternative network options is available in a first stage, and allocation takes place in a second stage after an optimal network is identified.

In the latter case, information about alternative network options makes it natural to apply the model of a cooperative game and invoke the standalone principle, defined in (1.8), as the basis of fairness. In the former case of fixed networks, however, this approach is often questionable, and ideas of fair allocation have to be related more directly to individual network usage.

2.1   Fixed Trees

We start out with fixed trees defined as (existing) cycle-free network structures where the cost (or revenue) of alternative structures are unknown or counterfactual. Unless explicitly stated, we assume connections (edges) can be used without rivalry (i.e., connections are public goods for the network users).

2.1.1   Chains

One of the simplest examples of a tree is a *linear hierarchy* representing a chain of command, an ownership, or recruitment structure. Here, a "boss" recruits a subordinate; this subordinate recruits an apprentice (who now becomes the subordinate of the subordinate of the boss), and so forth. Variations of this organizational structure can be found over a broad range, from law firms to social mobilization schemes (Hougaard et al., 2017). Other examples include the *river-sharing* problem, where different regions (or countries) share a water flow from a single source (Ambec and Sprumont, 2002), and the so called *airport model* arising from an empirical study of airport landing fees in Birmingham, UK (Littlechild and Thompson, 1977). Despite many structural similarities, these models are fundamentally different from an economic point of view. We will discuss the models and analyze possible solutions. In particular, we will see how the resulting allocations can be defined directly from the network structure itself, and alternatively, how the problem can be modeled as a cooperative game, where classic game-theoretic solution concepts may represent compelling ways to share payoff.

2.1.1.1   The linear hierarchy model

A linear hierarchy can be modeled as a simple chain-structured graph, where positions in the chain indicate a ranking of the agents (here identified by nodes). When each member of the hierarchy generates a personal revenue, the power structure of the hierarchy may justify revenue reallocation. We will show that a large family of so-called fixed fraction transfer rules can be characterized by a set of compelling normative requirements. For members of this family, application of the rule implies that revenue is "bubbling up" from subordinates to all their superiors.

Let $\mathcal{N}$ be the set of natural numbers identifying potential agents, and let $N \in \mathcal{N}$ be a finite set of such agents. Following the natural order of the numbers in $N = \{1, \ldots, n\}$ we let the largest number n denote the "boss" and lower numbers correspond to lower positions in the hierarchy (with 1 being the lowest-ranked member). In this sense, N represents a linear hierarchy: that is, a digraph with $D(n) = \emptyset$ and $D(i) = i + 1$ for $i = 1, \ldots, n - 1$.

Each agent $i \in N$ in the hierarchy generates a personal revenue $r_i \geq 0$ resulting in the revenue profile $r = \{r_i\}_{i \in N}$. However, due to the ownership (or power structure) behind the hierarchy, reallocation of individual revenues may be justified. For instance, part of the individual revenues can be transferred upward in the hierarchy, acknowledging the position of the boss and the power structure in general.

Formally, a (linear hierarchy) *revenue-sharing problem* is a pair (N, r) consisting of a linear hierarchy $N \in \mathcal{N}$ and a revenue profile $r \in \mathbb{R}^n_+$. Let $\mathcal{R}^N$ be the set of problems involving the hierarchy N and $\mathcal{R} = \cup_{N \in \mathcal{N}} \mathcal{R}^N$.

Given a problem $(N, r) \in \mathcal{R}$, a *revenue allocation* is a vector of payments $x \in \mathbb{R}^n_+$ satisfying *budget balance*, that is, $\sum_{i \in N} x_i = \sum_{i \in N} r_i$.

An *allocation rule* is a mapping ϕ assigning to each problem (N, r) an allocation $x = \phi(N, r)$. We assume that relevant allocation rules are *anonymous*, that is, for each problem (N, r) and for each strictly monotonic function $g : N \to N'$, we have $\phi_{g(i)}(N', r') = \phi_i(N, r)$, where $r'_{g(i)} = r_i$ for each $i \in N$. Since ϕ is anonymous, we can set $N = \{1, \ldots, n\}$ without loss of generality.

For $N = \{1, 2, 3\}$ the linear hierarchy can be illustrated by the following chain-structured digraph.

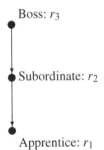

Boss: r_3

Subordinate: r_2

Apprentice: r_1

Even though the hierarchy is fixed in the short run, a regulator may still want to implement an allocation rule that justifies revenue reallocation in order to create stability over time and sustain future development of the organization. Potentially, multiple allocation rules can be applied by the regulator. Two polar examples are the following.

The no-transfer rule, ϕ^{NT} For each problem (N, r),

$$\phi^{NT} = r. \tag{2.1}$$

The full-transfer rule, ϕ^{FT}   For each problem (N, r),

$$\phi^{FT}(N, r) = \left(0, \ldots, 0, \sum_{i \in N} r_i\right). \tag{2.2}$$

Using the no-transfer rule, every agent keeps her personal revenue, while using the full-transfer rule, the boss gets the total revenue of the hierarchy and everybody else gets nothing. In between these two extremes we can imagine a vast number of alternative allocation rules where some part of the individual revenues are reallocated.

2.1.1.2 Axioms and characterization

To provide some structure on the set of relevant rules, we may apply the following set of fundamental incentive and fairness requirements (axioms).

Lowest-rank consistency   For $n \geq 2$, where $(N, r) \in \mathcal{R}$ and $(N \setminus \{1\}, (r_2 + r_1 - \phi_1(N, r), r_{N \setminus \{1,2\}})) \in \mathcal{R}$,

$$\phi_{N \setminus \{1\}}(N, r) = \phi\left(N \setminus \{1\}, (r_2 + r_1 - \phi_1(N, r), r_{N \setminus \{1,2\}})\right).$$

Consistency axioms generally refer to the way in which the rule reacts when agents leave the scene with their awarded payments. Lowest-rank consistency focuses on the agent with the lowest rank. It states that when this agent leaves, a new problem arises in which the immediate superior of agent 1 (i.e., agent 2) adds any remaining surplus from agent 1 to her revenue (which then becomes $r_2 + r_1 - \phi_1(N, r)$), and payments in this new problem should not differ from those in the original problem (for everyone but agent 1, who is out).

The next two properties focus on the opposite end of the hierarchy. The first says that the revenue generated by the highest-ranked agent (i.e., the boss) is irrelevant for the allocation of payments to all the subordinates. It seems reasonable when the boss is the indisputable owner of her own revenue.

Highest-rank revenue independence   For each $(N, r) \in \mathcal{R}$ and each $\hat{r}_n \geq 0$,

$$\phi_{N \setminus \{n\}}(N, r) = \phi_{N \setminus \{n\}}\left(N, (r_{N \setminus \{n\}}, \hat{r}_n)\right).$$

The second property avoids certain strategic manipulations of the allocation by the boss. If the boss splits her revenue into two amounts represented by two aliases, then the payoffs for all subordinates must be unaffected.

Highest-rank splitting neutrality   For each $(N, r) \in \mathcal{R}$, let $(N', r') \in \mathcal{R}$ be a new problem such that $N' = N \cup \{k\}$, $k > n$, $r_n = r'_k + r'_n$, and $r'_{N \setminus \{n, k\}} =$

$r_{N \setminus \{n\}}$. Then

$$\phi_{N \setminus \{n,k\}}(N', r') = \phi_{N \setminus \{n\}}(N, r).$$

Finally, it is natural to add a property stating that if revenues are scaled by a factor α, so are the payments.

Scale covariance   For each (N, r), and each $\alpha > 0$,

$$\phi(N, \alpha r) = \alpha \phi(N, r).$$

We are able to show that these properties jointly characterize a parametric family of rules, $\{\phi^\lambda\}_{\lambda \in [0,1]}$, which here are dubbed fixed-fraction transfer rules.[1] The family contains the no-transfer rule and the full-transfer rule as extreme members, corresponding to $\lambda = 1$, and $\lambda = 0$, respectively. Payments to individual agents, $i \in N$, are defined as

$$\phi_i^\lambda(N, r) = \begin{cases} \lambda(r_i + (1-\lambda)r_{i-1} + \cdots + (1-\lambda)^{i-1}r_1), & i = 1, \ldots, n-1, \\ r_i + (1-\lambda)r_{i-1} + \cdots + (1-\lambda)^{i-1}r_1, & i = n. \end{cases}$$

$$(2.3)$$

Using a fixed-fraction transfer rule, the agents keep a (fixed) fraction λ of their own revenue as well as any surplus "bubbling up" from the subordinates (and thereby agents transfer a fixed fraction $(1 - \lambda)$ upward in the hierarchy). The boss, n, is exceptional in the sense that she keeps her full revenue as well as any residual surplus from the entire hierarchy. In particular, note that the same fraction λ is used throughout the hierarchy. Formally, we state this as a theorem (see Hougaard et al., 2017).

Theorem 2.1   A rule ϕ satisfies lowest-rank consistency, highest-rank revenue independence, highest-rank splitting neutrality, and scale covariance if and only if it is a fixed-fraction transfer rule, that is, $\phi \in \{\phi^\lambda\}_{\lambda \in [0,1]}$.

Proof   Suppose that ϕ is a rule satisfying all the axioms in the statement of the theorem. First, let $N = \{1\}$ and $r = r_1$. By budget balance, $\phi_1(N, r) = r_1 = \phi_1^\lambda(N, r)$, for each $\lambda \in [0, 1]$. Next, add a superior agent 2 with revenue r_2. Let $N' = \{1, 2\}$ and $r' = (r_1, r_2)$. We claim that $\phi_1(N', r') \in [0, r_1]$, so $\phi_1(N', r') = \lambda r_1 = \phi_1^\lambda(N', r')$ for some $\lambda \in [0, 1]$. Suppose $\phi_1(N', r') > r_1$. Then by highest-rank revenue independence, $\phi_1(N', r') = \phi_1(N', (r_1, 0))$ so by budget balance, $\phi_1(N', r') \leq 1$—a contradiction.

---

1. In Hougaard et al. (2017) these rules are called *geometric* rules.

By highest-rank revenue independence, λ is independent of r_2. Moreover, λ is independent of r_1. Indeed, by contradiction, suppose that $\tilde{r} = (\tilde{r}_1, \tilde{r}_2)$ with $r_2 = \tilde{r}_2$, and $\phi_1(N', r') = \lambda r_1$, and $\phi_1(N', \tilde{r}) = \tilde{\lambda}\tilde{r}_1$ with $\lambda \neq \tilde{\lambda}$. Then, by scale covariance, $\phi_1(N', \frac{\tilde{r}_1}{r_1}r) = \frac{\tilde{r}_1}{r_1}\lambda r_1 = \lambda\tilde{r}_1 \neq \tilde{\lambda}\tilde{r}_1$, contradicting the assumption that λ is independent of r_2. Now, by budget balance, $\phi_2(N', r') = r_2 - r_1 - \phi_1(N', r') = \phi_2^\lambda(N', r')$.

Next, suppose there is λ such that $\phi = \phi^\lambda$ for all problems with up to k agents, $k \geq 2$. Now, consider the problem (N^k, r^k) with $N^k = \{1, \ldots, k\}$ and $r^k = \{r_1, \ldots, r_k\}$ and add an agent $k + 1$. By highest-rank revenue independence, and highest-rank splitting neutrality, $\phi_i(N^{k+1}, r^{k+1}) = \phi_i(N^k, r^k) = \phi_i^\lambda(N^k, r^k)$ for all $i \leq k - 1$. By lowest-rank consistency, $\phi_k(N^{k+1}, r^{k+1}) = \phi_k(N^{k+1} \setminus \{1\}, r_2^{k+1} + r_1^{k+1} - \phi_1(N^{k+1}, r^{k+1}), r_{N^{k+1}\setminus\{1,2\}})$, and thus, by the induction hypothesis, $\phi_k(N^{k+1}, r^{k+1}) = \phi_k^\lambda(N^{k+1}, r^{k+1})$. Finally, by budget balance, we have

$$\phi_{k+1}(N^{k+1}, r^{k+1}) = r_{k+1} - \sum_{i=1}^{k} \phi_i^\lambda(N^{k+1}, r^{k+1}) = \phi_{k+1}^\lambda(N^{k+1}, r^{k+1}).$$

The converse statement follows easily.   ∎

To get some intuition, consider the following example.

Example 2.1 (The mechanics of fixed-fraction transfer rules)   Consider three agents $N = \{1, 2, 3\}$. The lowest-ranked agent has revenue r_1, of which he keeps a fraction λ; thus the amount $(1 - \lambda)r_1$ bubbles up to his immediate superior, agent 2. Agent 2 keeps a fraction λ of her own revenue r_2 as well as of the part bubbling up from agent 1. Thus, the part $(1 - \lambda)^2 r_1$ of agent 1's revenue further bubbles up to the boss (agent 3) as well as the part $(1 - \lambda)r_2$ of agent 2's revenue r_2.

If only the lowest-ranked agent generates revenue, say, $r_1 = 1$ (with $r_2 = r_3 = 0$), and $\lambda = 0.5$, we therefore get payments

$$(x_1, x_2, x_3) = (0.5, 0.25, 0.25).$$

Generally, for any agent i, this can be written as

$$x_i = \frac{r_1}{2^i},$$

with the addition that the boss gets the residual revenue due to budget balance. Such a payment scheme has recently been analyzed in the context of social mobilization schemes (see Pickard et al., 2011, and the case description in the Introduction) where the hierarchy is interpreted as a chain of recruitment (agent

n calls agent $n - 1$, agent $n - 1$ calls agent $n - 2$, and so on). In this case teams were competing to locate a given object. The team finding the object wins a prize; the prize is shared using the above payment scheme (dubbed the MIT strategy after the winning team), and a team can keep calling new members until the object is found. In terms of incentives, it is claimed that the MIT strategy is superior to other payment schemes.

In our context it is interesting to note that if we let the hierarchy be endogenous, as in the MIT recruitment example, then λ may (in a myopic sense) be interpreted as the probability of getting a subordinate at any level in the hierarchy. Indeed, it seems natural to assume that the probability of recruiting a subordinate is closely connected to this subordinate's potential income. Hence, if $\lambda = 0$, no agent will join, since they transfer all their income to the superior; if $\lambda = 1$, agents will join with certainty, since they keep their full revenue.

Given that we restrict the boss to choose among fixed-fraction transfer rules, setting $\lambda = 0.5$ will, in fact, maximize the expected profit of the boss: Normalize all revenues to 1, then the boss has to solve $\max_\lambda \sum_{t=1}^{\infty}((1 - \lambda)\lambda)^t$ with solution $\lambda = 0.5$. Somewhat surprisingly, this remains true if agents are farsighted and take into account their ability to hire further subordinates from whom revenues will bubble up, in effect making the probability that they will join the hierarchy equal to their payoff from the rule (which is larger than λ). Indeed, let the probability δ equal the payoff, that is, $\delta = \lambda + \lambda \sum_{t=1}^{\infty}((1 - \lambda)\delta)^t$ when revenues are normalized. Then the boss has to solve $\sup_{\lambda,\delta} \sum_{t=1}^{\infty}((1 - \lambda)\delta)^t$. The equation in δ has two solutions: $\delta = 1$, $\lambda > 0$ and $\delta = \lambda/(1 - \lambda)$. In the first case, the solution is degenerate since it is profit maximizing for the boss to set $\lambda \to 0$. In the second case, it is profit maximizing to let $\lambda \to 0.5$ (in which case $\delta \to 1$).

To put the fixed-fraction transfer rules into perspective, consider other obvious candidates: for instance, a rule where all agents except the boss get a fixed salary, or a rule where agents are allowed to keep a fixed share λ of their individual revenue and the boss gets the residual. In the latter case, shares are given by $x_i = \lambda r_i$ for all $i = 1, \ldots, n - 1$ and $x_n = r_n + \sum_{i=1}^{n-1}(1 - \lambda)r_i$. Clearly, such rules are not members of the family of fixed-fraction transfer rules, $\{\phi^\lambda\}_{\lambda \in [0,1]}$, since no revenue is bubbling up from a subordinate to his superior. Technically speaking they violate lowest-rank consistency (while satisfying the remaining axioms). However, this does not indicate that lowest-rank consistency is tantamount to fixed-fraction transfer rules, since several rules that are quite far from the idea of revenue bubbling up satisfy lowest-rank consistency: for example, consider the rule where all superior agents, as well as agent i herself, share agent i's revenue equally (i.e., revenue shares are given by $x_i = \sum_{j=1}^{i} \frac{r_j}{n-j+1}$

for all i). This serial rule satisfies lowest-rank consistency, but compared to the family of fixed-fraction transfer rules, it violates highest-rank splitting neutrality.

The linear hierarchy revenue-sharing problem is an example of a network allocation problem that is difficult to give a natural representation as a cooperative game. Indeed, there are no externalities between agents. The hierarchy is the result of an ownership (or power) structure, and that is the only reason for potential reallocations of revenues. Also, the standalone revenues of all coalitions (except the grand coalition N) are counterfactual: the hierarchy is existing, and we are not in a design phase trying to optimize the size of the hierarchy. Consequently, coalitions' standalone values are not naturally specified.

Yet, in the literature, hierarchies have been interpreted as *permission structures*. Following this line of argument, the revenue-sharing problem can be imbedded in the framework of games with graph restrictions (see section 1.3.1), as we shall briefly describe below.

2.1.1.3 A game-theoretic-approach

Assume that the underlying economic situation can be described by an additive game in revenues, where the worth of a coalition is simply the sum of the members' individual revenues, that is, $v(S) = \sum_{i \in S} r_i$ for every coalition $S \subseteq N$ (reflecting the fact that there is no externality between agents). The linear hierarchy can be seen as imposing a permission structure that restricts this game, either in the sense that a given agent needs approval from all his superiors before he can cooperate (the *conjunctive* restriction) or that he needs approval from at least one of his superiors before he can cooperate (the *disjunctive* restriction).

The problem can then be construed as a *game with permission structure*.[2] In linear hierarchies the permission structure is so simple that the conjunctive and disjunctive restrictions coincide. The restriction on a given coalition S consists of members in S whose entire set of superiors are also part of S.

Let $i^{max}(T)$ denote the top-ranked agent in coalition T. Given the additive game v, then for any coalition $S \subseteq N$, the (conjunctive) restriction of the game v (denoted $\tilde{v}$) provided by the permission structure in the form of a linear hierarchy is given by

$$\tilde{v}(S) = v(S \setminus \{j \in N \mid j < i^{max}(N \setminus S)\}) = \sum_{h \in S \setminus \{j \in N \mid j < i^{max}(N \setminus S)\}} r_h. \qquad (2.4)$$

---

2. See Gilles et al. (1992) for analysis of the conjunctive restriction, and Gilles and Owen (1999) for analysis of the disjunctive restriction, on general permission structures.

Rewriting the value of any coalition $S \subseteq N$, we get

$$\tilde{v}(S) = \begin{cases} 0 & \text{if } i^{max}(S), < i^{max}(N \setminus S), \\ r_{i^{max}(N \setminus S)+1} + \cdots + r_n & \text{otherwise.} \end{cases} \qquad (2.5)$$

That is, the restricted value of coalition S is 0 unless S includes an upper segment containing the boss: in this case the restricted value is found by sequentially adding the revenues from the boss and downward until an agent from the complement appears in the hierarchy.

Revenues can be allocated according to some game-theoretic solution concept with respect to the permission-restricted game $\tilde{v}$: for instance, the Shapley value. Here the Shapley value splits a given agent's revenue equally among this agent and all his superiors. The payoff to a given agent i is therefore given by

$$\phi_i^{Sh}(N, \tilde{v}) = \sum_{j=1}^{i} \frac{r_j}{n - j + 1}. \qquad (2.6)$$

As already noted, (2.6) violates the axiom of highest-rank splitting neutrality (so it can be manipulated by the boss creating multiple aliases). It is radically different from the parametric family of fixed-fraction transfer rules analyzed above and is less compelling in the present case where there are no obvious externalities between agents.

Axiomatic characterizations of the *permission value* (i.e., the Shapley value of the permission-restricted game) can be found in van den Brink and Gilles (1996) and van den Brink (1997).

2.1.1.4 River sharing

Ambec and Sprumont (2002) present a model of a similar type of (fixed) chain structured network, but this time with externalities. A set of regions (countries) N are ordered by the water flow in a river with a single source. Water flows from the source (considered as region 1) and downstream to the regions ordered as $2, 3, \ldots, n$. Thus, if $i < j$, then i is upstream from j. The flow at the source is given by $f_1 > 0$, and for every downstream region i, the flow is increased by the (region controlled) water flow $f_i > 0$. Due to the directed flow, water added at region i, can only be consumed by i and its downstream regions. Thus, the total amount of water ending up at region n is $\sum_{i \in N} f_i$ if no water is consumed along the river.

Regions benefit from water consumption as given by a region-specific benefit function $b_i(x_i)$, where x_i is the amount of water consumed by region i. It is assumed that every function b_i is strictly increasing; strictly concave; and differentiable for $x_i > 0$, with its derivative going to infinity as x_i tends to zero.

A consumption plan $x = (x_1, \ldots, x_n)$, for N, is *feasible* if

$$\sum_{i \leq j} x_i \leq \sum_{i \leq j} f_i, \quad \text{for all} \quad j \in N. \tag{2.7}$$

A feasible consumption plan x^* is *welfare maximizing* if it solves

$$v(N) = \max_x \sum_{i \in N} b_i(x_i). \tag{2.8}$$

By the assumptions on the benefit functions b_i, x^* is uniquely determined, and the resulting distribution of welfare is given by $(b_1(x_1^*), \ldots, b_n(x_n^*))$.

However, there is no reason to expect that a welfare-maximizing water allocation is particularly fair. For instance, consider the case where water is added only at the source (i.e., $f_1 > 0$ and $f_i = 0$ for $i \neq 1$) and all regions have the same benefit function $b = b_i$ for all i. In this case a welfare-maximizing consumption plan will split region 1's water (f_1) equally: $x^* = (\frac{f_1}{n}, \ldots, \frac{f_1}{n})$. However, region 1 should arguably have the right to consume all the water it controls, which would generate the benefit $b(f_1) > b(\frac{f_1}{n})$, since b is strictly increasing. This seems to call for compensation in the form of side payments.

Indeed, assuming the regions have quasilinear preferences over water and money, any compelling welfare allocation $z \left(\sum_{i \in N} z_i = v(N)\right)$ can be obtained by side payments t_i determined by $t_i = z_i - b_i(x_i^*)$, for every region $i \in N$.

Due to the presence of group externalities, it becomes natural to model the problem as a cooperative game. So we need to determine the consumption plan that maximizes welfare for every coalition of regions $S \subset N$. But what is feasible for S depends on the complement's water consumption: the worst case for S is that $N \setminus S$ consumes all the water it controls; the best case is that $N \setminus S$ consumes nothing.

Formally, for every coalition $S \subset N$ there is a unique coarsest partition $\mathcal{S}$ of consecutive components T (T is consecutive if $k \in T$ when $i, j \in T$ and $i < k < j$). Thus, in the worst case, a consumption plan x for coalition S is feasible if

$$\sum_{\{i \leq j\} \cap T} x_i \leq \sum_{\{i \leq j\} \cap T} f_i \quad \text{for all} \quad j \in T \quad \text{and} \quad T \in \mathcal{S}. \tag{2.9}$$

In the best case, a consumption plan x for coalition S is feasible if

$$\sum_{i \in \{j \leq i\} \cap S} x_i \leq \sum_{i \in \{i \leq j\}} f_i, \quad \text{for every} \quad j \in S. \tag{2.10}$$

Denote by $v(S)$ and $w(S)$ the maximal welfare of coalition S under (2.9) and (2.10), respectively (again the assumptions on b_i ensure a unique solution in both cases). As argued by Ambec and Sprumont (2002), $v(S)$ is a natural lower bound on the welfare that coalition S can rightfully claim, while $w(S)$ is a natural upper bound. They prove that there is a unique allocation rule satisfying both these bounds, dubbed the *downstream incremental solution*:

$$z_i^{DI} = v(\{j \le i\}) - v(\{j < i\}), \quad \text{for all} \quad i \in N. \tag{2.11}$$

Since $v(\{i \le j\}) = w(\{i \le j\}$, for every j, we can equivalently define the downstream incremental solution with respect to the game (N, w), that is, as $z_i^{DI} = w(\{j \le i\}) - w(\{j < i\})$, for all $i \in N$. So the games (N, v) and (N, w) are balanced, since z^{DI} is a core solution in both (in fact, the unique intersection).

It is not straightforward to relate the family of fixed-fraction transfer rules (2.3) to the river-sharing problem, and because of the (group) externalities involved here, it may not be as compelling as in the linear hierarchy context either.

The river-sharing model has been generalized in various ways, including: allowing for satiation, multiple springs (sources), and uncertainty of agent-controlled water flows f_i (see Ambec and Ehlers, 2008; van den Brink et al., 2012; Ambec et al., 2013).

Next we focus on a similar type of model, but only from the cost side (i.e., as a pure cost-allocation problem). This has the obvious advantage that we avoid dealing with agent-specific benefits: they are typically private information of the agents and so are problematic to compare.

2.1.1.5 The airport model
In a cost-sharing context, the *airport problem* is a classic example of an allocation problem with a somewhat similar type of chain structure. As in the river-sharing problem, there are externalities involved (now in the form of cost externalities), and a representation of the problem as a cooperative game seems natural.[3]

The costs of an airport system of runways has a particularly simple structure, since the cost of building a runway is determined by the largest aircraft for which it is designed, and the cost of using the runway is proportional to the number of landings or takeoffs for each type of aircraft. We focus on the allocation of runway construction costs between the different aircraft types.

---

3. See, e.g., Littlechild and Owen (1973), Littlechild (1974), Littlechild and Thompson (1977), or the survey by Thomson (2013).

Let $N = \{1, \ldots, n\}$ be a (finite) set of n different aircraft types. Each type $i \in N$ is characterized by the standalone cost C_i, which is the cost of building a runway designed for type i. Larger aircraft require longer, and thus more expensive, runways. Hence the airplanes can be ordered and labeled according to increasing standalone cost,

$$0 \leq C_1 \leq \cdots \leq C_n.$$

Let $C = \{C_i\}_{i \in N}$ denote the profile of increasing standalone costs. An *airport problem* is a pair (N, C). Given a problem (N, C), an allocation is a vector of payments $x \in \mathbb{R}^n$ satisfying budget balance (i.e., $\sum_{i \in N} x_i = C_n$) and *boundedness* (i.e., $0 \leq x_i \leq C_i$). An *allocation rule* is a mapping ϕ assigning to each problem (N, C) an allocation $x = \phi(N, C)$.

The graph below illustrates the chain structure when $0 < C_1 < \cdots < C_n$.

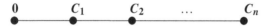

The cost of the first edge (from 0 to 1) is C_1, the cost of the second edge (from 1 to 2) is equal to the incremental cost $C_2 - C_1$, and so forth. The total cost of the runway is C_n (which has to be allocated).

2.1.1.6   Cost allocation in chains

When it comes to allocation, the linear hierarchy problem and the airport problem are quite different despite the obvious structural similarities. The main difference is caused by the presence of (cost) externalities in the airport problem. Therefore, the following allocation rule is very natural in the airport problem, in contrast to the linear hierarchy problem: Since all aircraft use the first segment of the runway (the first edge in the graph) all of them should share the cost, C_1, of that segment equally. Thus, all aircraft are allocated the cost share $\frac{C_1}{n}$ for the first segment of the runway. Since all aircraft except for the smallest (type 1) use the second segment of the runway as well (second edge in the graph), aircraft types $\{2, \ldots, n\}$ should share the incremental cost, $C_2 - C_1$, equally. Thus, aircraft types $\{2, \ldots, n\}$ are further allocated $\frac{C_2 - C_1}{n-1}$ for the second segment of the runway, and so forth, until the last incremental part of the runway (i.e., the edge from $n - 1$ to n with cost $C_n - C_{n-1}$) is paid solely by the largest aircraft type n.

Consequently, the cost share x_i of aircraft type $i \in N$ is determined as

$$x_i = \sum_{j=1}^{i} \frac{C_j - C_{j-1}}{n - j + 1}. \tag{2.12}$$

This allocation principle is known as the *serial principle* to cost sharing (see, e.g., Hougaard, 2009). As a cost-allocation rule it appears under many different names in the literature (e.g., the *sequential equal-contributions rule*; Thomson, 2013).[4]

The serial reasoning combines some form of egalitarianism (everybody shares equally the costs of the segments they use) with some kind of responsibility for own demand (no agent pays for segments that she is not using).

Instead of equal sharing among segment users, each type can also be made fully responsible for the segment added because of their specific presence. That is, each aircraft type $i \in N$ pays according to its incremental cost,

$$x_i = C_i - C_{i-1}. \tag{2.13}$$

This allocation principle is known as the *incremental principle* to cost sharing (see, e.g., Hougaard, 2009). In the graph, this corresponds to each agent paying the cost of the edge leaving the agent's node on the unique connected path to the root (here, the start of the runway). Such allocations are called Bird allocations in the following (after Bird, 1976). As a cost-allocation rule it appears under many different names in the literature (e.g., the *sequential full-contributions rule*; Thomson, 2013).

The fairness of Bird allocations may seem questionable, since some aircraft types may contribute very little if their incremental cost is small compared to the cost of building a runway for this type alone (the standalone cost). For example, if $C_1 = C$ and $C_2 = C + \varepsilon$, where $\varepsilon > 0$ is small, the Bird allocation becomes $(x_1, x_2) = (C, \varepsilon)$. In this case, type 1 pays almost the entire construction cost, while type 2 basically free rides, even though type 2 has the highest standalone cost. Clearly, this is not the case using the sequential equal-contributions rule (given by 2.12), where the resulting cost allocation is $(x_1, x_2) = (C/2, C/2 + \varepsilon)$. As we shall see in 2.1.1.8, however, other features speak in favor of (versions of) the sequential *full*-contributions rule when an ordering of the agents based on standalone costs seems less relevant for the allocation problem.

Alternatively, we may also use allocation rules that do not relate directly to the chain structure, for example, sharing in proportion to standalone cost C_i:

$$x_i = \frac{C_i}{\sum_{j \in N} C_j} \, C_n \quad \text{for all } i \in N. \tag{2.14}$$

---

4. Note the connection to the Shapley value of the revenue-sharing game restricted by the conjunctive permission structure of the linear hierarchy, given by (2.6).

However, using cost shares given by (2.14) makes the payment of a given type depend on the standalone cost of aircraft types that requires longer runways than the type in question. It can be argued that the cost share of an aircraft type—say, i—should be independent of standalone costs of aircraft types $j > i$, as we will discuss further in the following.

Clearly, several other allocation rules can be applied in this case. The best-known rules are discussed and analyzed in, for example, Thomson (2013) and Hougaard (2009).

2.1.1.7   The airport problem as a game

Instead of using allocation methods that relate directly to the structure of the cost graph, the airport problem may also be construed as a cost-sharing game, modeling the cost of every coalition of aircraft types.[5]

Since a runway of a certain length can serve aircraft requiring that length or shorter, the costs of building a runway that can serve an arbitrary group of aircraft equal that of serving the largest aircraft in the group. Thus, for every coalition of aircraft types, $S \subseteq N$, the cost of serving S is given by

$$c(S) = \max_{i \in S}\{C_i\}, \quad \text{with} \quad c(\emptyset) = 0. \tag{2.15}$$

The pair (N, c), where N is the set of agents (aircraft) and c is given by the cost function (2.15), will be called an *airport game*.

Since $c(S \cup T) = \max\{c(S), c(T)\}$ and $c(S \cap T) \leq \min\{c(S), c(T)\}$ we have that for every $S, T \in N$, $c(S \cup T) + c(S \cap T) \leq c(S) + c(T)$, so the airport game is *concave*.

Note further that for airport problems, the entire cost structure is easily determined in practice, since we only need to establish the standalone cost, $c(\{i\}) = C_i$, for every aircraft type i, in order to determine the entire game. This stands in contrast to the general TU-game model, where we need to establish the cost of every one of $2^n - 1$ coalitions (see section 1.2).

It turns out that this particular structure of the cost function further simplifies the computation of the well-known solution concepts, as shown for the nucleolus in Littlechild (1974) and for the Shapley value in Littlechild and Owen (1973). In fact, the Shapley value of the airport problem (N, c) coincides with the sequential equal-contributions rule, where cost shares are given by (2.12). Note, for example, that the marginal costs of agent i, $c(S \cup i) - c(S)$, differs from zero only when $C_i > \max_{j \in S}\{C_j\}$. Hence, the

---

5. Note that this is a more direct approach to establishing the game than the approach in the linear hierarchy case (see section 2.1.1.1), where the graph (hierarchy) was restricting an underlying additive game (in revenues).

Shapley value of type 1 is $\frac{0!(n-1)!}{n!}C_1 = \frac{1}{n}C_1$, and the Shapley value of type 2 is $\frac{0!(n-1)!}{n!}C_2 + \frac{1!(n-2)!}{n!}(C_2 - C_1) = \frac{1}{n}C_1 + \frac{1}{n-1}(C_2 - C_1)$, and so forth. So in the airport game, the Shapley value has a natural interpretation along the lines of the serial cost-sharing principle.

Since the airport game is concave, the core is rather large and contains both the nucleolus and the Shapley value. As such, the Shapley value ensures that no coalition of aircraft types pays more than their standalone costs, providing another strong argument in favor of the Shapley value.

Moreover, since the game is concave, there is a unique Lorenz-undominated allocation[6] in the core of the game (Dutta and Ray, 1989) often called the *Lorenz solution*. The Lorenz solution is the most equally distributed allocation in the core, so choosing the concept of Lorenz solution for the airport game obviously relies on a very direct form of egalitarianism. It may nevertheless result in somewhat unfair distributions of costs, as shown by the following simple example.

Example 2.2 (The Lorenz solution) Consider a situation with three aircraft types $N = \{1, 2, 3\}$, and the following cost structure: $(C_1, C_2, C_3) = (1, 2, 3)$.

The core the game (N, c) is given by allocations x satisfying $x_1 + x_2 + x_3 = 3$, where

$$0 \leq x_1 \leq 1,$$

$$0 \leq x_2 \leq 2,$$

$$1 \leq x_3 \leq 3.$$

Therefore, the Lorenz-maximizing element in the core is given by the equal split $(x_1, x_2, x_3) = (1, 1, 1)$. Yet an equal split does not seem particularly compelling in this case. For instance, aircraft type 1 pays its standalone cost, while both aircraft types 2, and 3 gain compared to their standalone cost. So given the restrictions of the core, aircraft type 1 pays its maximum, while aircraft type 3, pays its minimum.

As mentioned above, the nucleolus is particularly simple to compute in the airport game and is designed to maximize the minimal excess of the various coalitions, as shown in section 1.2.2. In the case $0 < C_1 < \cdots < C_n$, the

---

6. For two increasingly ordered allocations $x, y \in \mathbb{R}^n$ where $\sum_{i=1}^{n} x_i = \sum_{i=1}^{n} y_i$ we say that x *Lorenz dominates* y if $\sum_{i=1}^{k} x_i \geq \sum_{i=1}^{k} y_i$ for all $k = 1, \ldots, n - 1$. If x Lorenz dominates y, then payments in allocation x are more equally distributed than payments in allocation y.

nucleolus can be written as the allocation

$$x_1^{Nuc} = \min \left\{ \frac{C_1}{2}, \frac{C_2}{3}, \frac{C_3}{4}, \ldots, \frac{C_{n-1}}{n} \right\},$$

$$x_2^{Nuc} = \min \left\{ \frac{C_2 - x_1}{2}, \frac{C_3 - x_1}{3}, \ldots, \frac{C_{n-1} - x_1}{n-1} \right\},$$

$$x_3^{Nuc} = \min \left\{ \frac{C_3 - x_1 - x_2}{2}, \ldots, \frac{C_{n-1} - x_1 - x_2}{n-2} \right\},$$

$$\vdots$$

$$x_{n-1}^{Nuc} = \min \left\{ \frac{C_{n-1} - x_1 - \cdots - x_{n-2}}{2} \right\},$$

$$x_n^{Nuc} = C_n - x_1^{Nuc} - \cdots - x_{n-1}^{Nuc}.$$

Example 2.2 (continued: The nucleolus) Using this definition, we can compute the nucleolus of the game in example 2.2, yielding the allocation

$$x^{Nuc} = (0.5, 0.75, 1.75).$$

It can be checked that this allocation satisfies the core conditions (recall the nucleolus is a core-allocation rule). Compared to the equal-split allocation (in this case, the Lorenz maximal element of the core), the nucleolus seems more compelling, since individual excesses, $c(\{i\}) - x_i$, are more equally distributed. However, note that in general the cost shares of the nucleolus do not make the cost share of a given aircraft type independent of the stand-alone costs of aircrafts requiring longer (and more expensive) runways. This seems unfortunate, since aircraft i does not gain from the presence of aircraft j when $i < j$.

2.1.1.8 Axiomatic characterizations

In the airport game, the Shapley value can be given two simple axiomatic characterizations, demonstrating that the Shapley value is, in many ways, a natural choice of allocation rule when the cost-allocation problem can be represented as an airport game with cost function (2.15).

Consider the independence property mentioned in example 2.2 (continued) above, stating that the cost share of a given aircraft type $i \in N$, should not depend on the standalone cost of downstream agents (or, in other words, the segments of the runway that type i does not need to use). One version of such a property can be defined formally as follows.

Independence of at-least-as-large costs Let C and C' be two cost profiles, where $C_i' = C_i$. If for each $j \in N \setminus \{i\}$ such that $C_j < C_i$, we have $C_j' = C_j$ and for each $j \in N \setminus \{i\}$ such that $C_j \geq C_i$, we have $C_j' \geq C_j$, then $\phi_i(N, c') = \phi_i(N, c)$.

Moreover, it seems natural to require that if two aircraft types have the same standalone cost (i.e., need the same length of runway) they should pay the same. Formally, we have the following.

Equal-treatment-of-equals If $C_i = C_j$ then $\phi_i(N, c) = \phi_j(N, c)$.

As shown in Moulin and Shenker (1992) and Potters and Sudhölter (1999), these two properties uniquely characterize the Shapley value.

Theorem 2.2 Consider airport games (N, c). An allocation rule ϕ satisfies independence of at-least-as-large costs and equal-treatment-of-equals if and only if ϕ is the Shapley value.

Proof Let $J = |\{j \mid C_j < C_n\}|$. By induction on J; if $J = 0$, equal-treatment-of-equals and budget balance implies that $\phi_i(N, c) = \frac{C_n}{n} = \phi_i^{Sh}(N, c)$ for all $i \in N$. Assume the theorem holds for $J < k$. For $J = k$, we have $C_1 \leq C_2 \leq \cdots \leq C_k < C_{k+1} = \cdots = C_n$. Define a new cost profile by $C_i' = \min\{C_i, C_k\}$. By the induction hypothesis, $\phi(N, c') = \phi^{Sh}(N, c')$. By independence of at-least-as-large costs, $\phi_i(N, c) = \phi_i(N, c') = \phi_i^{Sh}(N, c)$ for $i \leq k$. By equal-treatment-of-equals, $\phi_i(N, c) = \phi_n(N, c)$, and $\phi_i^{Sh}(N, c) = \phi_n^{Sh}(N, c)$ for $i > k$, where $\phi_n(N, c) = \phi_n^{Sh}(N, c)$ by budget balance. ∎

Remark Note that the allocation rule where each type i pays its incremental cost, but if two or more types have the same standalone cost they share the incremental cost equally, violates independence of at-least-as-large costs. For instance, let $C_1 = C_2 = C$ and $C_1' = C_1, C_2' > C_2$. By independence of at-least-as-large costs, type 1 should pay the same in both cases, but we get $\phi_1(c) = \frac{C}{2}$ and $\phi_1(c') = C$. Such a rule satisfies a weaker version that refers to independence of *greater* costs (replacing weak with strict inequality in the last part of the definition of independence of at-least-as-large costs).

Another property that may seem reasonable is related to independence of the way that costs are categorized. If two cost profiles are ordered in the same way (e.g., if two runways were built at the same facility), then it seems natural to require that cost shares of the added problem should be the same as adding up the cost shares of each problem considered separately. The formal definition is as follows.

Conditional cost additivity   Let C' and C be two cost profiles that are ordered in the same way, that is, $C_1 \leq \cdots \leq C_n$ and $C'_1 \leq \cdots \leq C'_n$. Then $\phi_i(N, c' + c) = \phi_i(N, c') + \phi_i(N, c)$, for all $i \in N$.

If two similarly ordered cost profiles C and C' are added up, the resulting cost profile will be of the same order, and budget balance is preserved in the sense that $C_n + C'_n = (C + C')_n$ (note that this may not be the case if the two profiles were not similarly ordered). As shown in Dubey (1982), the conditional cost additivity property, together with equal treatment of equals, uniquely characterizes the Shapley value.

Theorem 2.3   Consider airport games (N, c). An allocation rule ϕ satisfies conditional cost additivity and equal-treatment-of-equals if and only if ϕ is the Shapley value.

Proof [Sketch of argument]   If $C_1 = \cdots = C_n$ then by equal-treatment-of-equals and budget balance, $\phi_i(N, c) = \frac{C_n}{n} = \phi_i^{Sh}(N, c)$ for all $i \in N$. Let $C_1 = \cdots = C_{n-1} < C_n$, and note that $C = C' + C''$, where $(C'_1, \ldots, C'_n) = (C_{n-1}, \ldots, C_{n-1})$ and $(C''_1, \ldots, C''_n) = (0, \ldots, 0, C_n - C_{n-1})$. Hence, by conditional cost additivity and budget balance, we get $\phi_i(N, c) = \phi_i(N, c') + \phi_i(N, lc'') = \frac{C_{n-1}}{n} + 0 = \phi_i^{Sh}(N, c)$ for $i \neq n$, and $\phi_n(N, c) = \phi_n(N, c') + \phi_n(N, c'') = \frac{C_{n-1}}{n} + C_n - C_{n-1} = \phi_n^{Sh}(N, c)$, and so forth.   ∎

Alternative characterizations of both the Shapley value and variations of the nucleolus can be found, for example, in Aadland and Kolpin (1998)[7] and Potters and Sudhölter (1999).

When there is no downstream cost externality between agents, the independence argument (as in independence of at-least-as-large costs) can be taken a step further. At least in some cases, it can be argued that not only should the cost share of type i be independent of the standalone costs of downstream agents j (with $C_j > C_i$), but it should also be independent of the *number* of such agents. This may seem unreasonable in the airport problem, but imagine, for example, that the chain represents a single data flow shared by several receivers as in a client-server network. In this case sharing the cost of establishing the connection (or maintenance costs) independent of the number of downstream clients may seem reasonable indeed (but if it concerns costs in the form of congestion, then clearly it is not).

---

7. The paper also contains an empirical study related to maintenance costs of common irrigation ditches in Montana. Interestingly, two cost-allocation rules have emerged in practice and have been in use since the early 1900s: the serial rule, and proportional cost sharing with respect to farm size or water share.

If all agents should be fully independent of downstream agents, this becomes a very strong condition that essentially singles out the Bird allocation defined in (2.13). Indeed, consider first the case where $N = \{1\}$. By budget balance $x_1 = C_1$. Now add another agent with $C_2 > C_1$. By full independence of downstream agents $x_1 = C_1$, and by budget balance $x_2 = C_2 - C_1$, and so forth, repeating the use of the full-independence assumption.

A weaker version of this property plays a central role in a characterization of the parametric family of fixed-fraction transfer rules (2.3) in the context of airport problems (N, C). The family of rules obtained by replacing r_i with $C_i - C_{i-1}$ (with the convention that $C_0 = 0$) in the definition (2.3) is, in the cost-sharing context, dubbed the family of *sequential λ-contributions rules*.

Consider a variable populations setting and assume that $0 < C_1 < \cdots < C_n$. The following property states that, if an agent does not have the largest stand-alone cost in the group, the cost share should depend only on the costs of the (unique connected) path from the agent to the root. The formal definition is as follows.

Independence of downstream agents   For two sets of agents N and N' with $\{1, \ldots, m+1\} \subset N \cap N'$ and cost profiles C and C' with $C_i = C_i'$ for all $i \leq m$,

$$\phi_i(N, C) = \phi_i(N', C')$$

for all $i \leq m$.

In other words, if two cost-allocation problems coincide for a lower coalition of agents with no agent being the last agent, then these agents should pay the same amount. This property is clearly satisfied by the sequential full-contributions rule, but it is *not* satisfied by the sequential equal-contributions rule that depends on the total number of agents in the chain. Notice that independence of downstream agents implies highest rank revenue independence and highest rank splitting neutrality (with the straightforward reformulation to the context of the airport problem).

The equivalent to lowest rank consistency in the linear hierarchy model is here called first-agent consistency (see Potters and Sudhölter, 1999): For $n \geq 2$, consider the cost profile C, and let $\phi(N, C)$ be the associated cost allocation. Then excluding agent 1 results in a new cost profile C_{-1}^{ϕ} for the reduced problem where, agent 1's cost share, $\phi_1(N, C)$, is subtracted from the standalone cost of all the remaining agents:

$$C_{-1}^{\phi} = (C_2 - \phi_1(N, C), \ldots, C_n - \phi_1(N, C)).$$

We now require that the allocation among agents $\{2, \ldots, n\}$ is unchanged by the exclusion of agent 1, defined formally as follows.

First-agent consistency   For all chain problems (N, C),

$$\phi_i(N \setminus \{1\}, C^\phi_{-1}) = \phi_i(N, C)$$

for all $i \geq 2$.

Potters and Sudhölter (1999) show that several rules satisfy first-agent consistency, for example, both the sequential equal- and the sequential full-contributions rule, and by their theorem 2.4, all rules satisfying first-agent consistency result in core allocations in the airport game (2.15).

Now, adding the property of **scale covariance** (i.e., $\phi(N, \alpha C) = \alpha \phi(N, C)$ for all factors $\alpha > 0$) the equivalent of theorem 2.1 for airport problems becomes the following.

Corollary 2.3.1   Consider the set of airport problems (N, C). A cost allocation rule ϕ satisfies independence of downstream agents, first-agent consistency, and scale covariance if and only if $\phi(N, C) = \phi^\lambda(N, C)$, defined in (2.3) by replacing r_i with $C_i - C_{i-1}$ (with $C_0 = 0$).

The family of sequential λ-contributions rules, ϕ^λ, encompasses all rules where the first $n - 1$ agents pay a share $\lambda \in [0, 1]$ of their incremental cost as well as any remaining debt from prior agents, and the last agent n pays the residual. Hence, cost shares can be determined recursively as $x_1^\lambda = \lambda C_1$ for agent 1, $x_2^\lambda = \lambda(C_2 - C_1 + (1 - \lambda)C_1) = \lambda(C_2 - x_1^\lambda)$ for agent 2, and so forth. Clearly, ϕ^1 is the sequential full-contributions rule yielding the allocation defined in (2.13). The opposite extreme, ϕ^0, represents the case where the first $n - 1$ agents free ride and the last agent pays the total cost.

The sequential full-contributions rule ϕ^1 can be singled out by replacing first-agent consistency and scale covariance with a property stating that cost shares shall be weakly increasing in edge costs: (that is, if $C_j - C_{j-1} \geq C_k - C_{k-1}$, then $\phi_j(N, C) \geq \phi_k(N, C)$. The argument is straightforward. Let $N = \{1\}$ and $C = C_1$. By budget balance, $\phi_1(N, C) = C_1 = \phi_1^1(N, C)$. Next, add a second agent with $C_2 > C_1$. Let $N' = \{1, 2\}$ and $C' = (C_1, C_2)$. Then $\phi_1(N', C') \in [0, C_1]$, so $\phi_1(N', C') = \lambda C_1 = \phi_1^\lambda(N', C')$ for some $\lambda \in [0, 1]$. By independence of downstream agents, λ is independent of C_2. Now, assume that $\lambda < 1$. Then by budget balance, we may choose C_2 (e.g., as $C_2 = C_1 + \lambda C_1$) such that cost shares are nonincreasing in edge costs. Thus, $\lambda = 1$ as desired.

Example 2.3 (Illustrating the λ-contributions rule)   Consider a client-server network: say, the server is a storage facility that clients connect to in a chain. Suppose three clients $N = \{1, 2, 3\}$ are located as in the graph below. Network maintenance costs are given by $k_i > 0$. In a client-server network, the order of the agents does not have any specific interpretation, since it may be determined by design or by evolution of the network. Therefore it is not obvious that externalities should play a role here.

Using a λ-contributions rule, client 1 pays a share λ of k_1; client 2 pay a share λ of k_2 and the uncovered cost $(1-\lambda)k_1$; client 3 pays the residual. That is,

$$x_1 = \lambda k_1,$$

$$x_2 = \lambda(k_2 + k_1 - x_1) = \lambda(k_2 + (1-\lambda)k_1),$$

$$x_3 = k_1 + k_2 + k_3 - x_1 - x_2 = k_3 + (1-\lambda)k_2 + (1-\lambda)^2 k_1.$$

Notice that this differs from another straightforward generalization of the Bird allocation, where agents pay a share α of the cost of their exiting edge, and the last agent pays the residual:

$$x_1 = \alpha k_1,$$

$$x_2 = \alpha k_2,$$

$$x_3 = k_3 + (1-\alpha)k_2 + (1-\alpha)k_1.$$

This rule violates first-agent consistency (while satisfying the other three properties of corollary 2.3.1).

We may also interpret a chain-structured graph as a *time line*. Abusing notation, we can consider the increment $C_i - C_{i-1}$ as the cost of the time span assigned to agent i, now identifying agents with edges. For instance, $[0, C_n]$ can be the total cost of the full range of opening hours of a given facility, and $[C_{i-1}, C_i[$ can be the cost of the time slot assigned to agent i. Consider the example below.

Example 2.4 (A time line interpretation) Consider three airline companies, A, B, and C, sharing a gate that is open for 12 hours. Airline A is assigned time slot $[0, 5[$, B is assigned time slot $[5, 7[$, and C time slot $[7, 12]$. Customers of all airlines can use gate facilities at any time during open hours. Thus, an airline partially benefits from open hours prior to its own time slot, since its customers have access to the gate facilities. The airport has to allocate the total cost of having the gate open for 12 hours among the three airline companies. The figure below illustrates the situation.

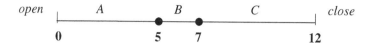

open |———————A———————B———————C———————| *close*
0   5   7   12

Assume that the cost of having the gate open is 1 monetary unit per hour, so the three airlines face the problem of sharing the total cost of 12.

Using the Bird rule, we get $(x_A, x_B, x_C) = (5, 2, 5)$. Using the serial rule (2.12), we get $(x_A, x_B, x_C) = (5/3, 8/3, 23/3)$. Clearly, the Bird rule ignores the benefit from customers being able to use the gate facilities prior to their own departure. But the serial rule fails to reflect the length of the time slot used by each company.

The sequential λ-contribution rule represents a kind of compromise with the Bird rule as one extreme case. Consider $\lambda = 0.5$, which yields the cost shares $(x_A, x_B, x_C) = (2.5, 2.25, 7.25)$. Airline B now pays less than airline A and yet more than its incremental cost. Note that airline C, being the last and largest, naturally pays the most.

2.1.2 Standard Fixed Trees

Generally, we may consider any kind of rooted tree structure as in the standard fixed-tree model.[8] For instance, the tree can represent a multicast data flow, where edge costs can be measures of usage or maintenance of network facilities.

A standard fixed tree T is a weighted rooted tree where each node is occupied by one or more agents, and each edge ij is weighted by its cost k_{ij}. For simplicity we assume that only one edge leaves the root. An example of a fixed rooted tree is illustrated below in the specific case where five agents (indexed by bold numbers) occupy one node each and weights are given by edge costs k_{ij}.

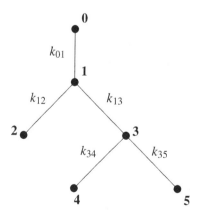

---

8. See, e.g., Megiddo (1978a), Granot et al. (1996), Herzog et al. (1997), Koster et al. (2001), and Bjørndal et al. (2004).

Clearly the chain representing the airport game is a special case of a standard fixed tree where all agents want connection to the root and edge costs are given by $k_{ij} = C_j - C_i$ for adjacent nodes $j > i$.

2.1.2.1 The model and some allocation rules

A *fixed-tree problem* is a tuple (N^0, E, k), where $N^0 = N \cup \{0\}$ is the set of agents including the root 0, E is the set of edges in the tree, and k is a vector of edge costs. All agents want connection to the root, and assume for simplicity that only one agent is present at each node in the tree. Moreover, simplifying the notation, we let k_i denote the edge cost of agent i (i.e., the cost of the edge leaving node i on agent i's unique connected path to the root). The total cost of the tree is therefore $C = \sum_{i \in N} k_i$.

A cost-allocation rule ϕ assigns a vector of cost shares $x = \phi(N^0, E, k) \in \mathbb{R}^N$ to each problem (N^0, E, k) satisfying budget balance: $\sum_{i \in N} x_i = C$.

Assuming there is no externality among agents, some form of independence of downstream agents seems reasonable (e.g., independence of edge costs associated with downstream agents). Thus, the cost share of an agent at node j in the tree should be independent of edge costs related to agents for whom the unique path to the root includes the node j. Consequently, it can be argued that the serial principle applies naturally in this more general setting as well. That is, the cost of each edge should be split equally among all downstream agents.

Formally, consider a given problem (N^0, E, k). Let p^i denote the set of nodes on agent i's path to the root (including node i itself). Moreover, let $n(i)$ be the set of agent i's downstream nodes (including node i itself), that is, agents j for which $p^j \ni i$. Now, according to the serial rule, every agent i gets a cost share

$$x_i^S = \sum_{j \in p^i} \frac{1}{|n(j)|} k_j. \tag{2.16}$$

Loosely speaking, if we extend the independence argument to include the *number* of downstream agents, we will again arrive at the Bird allocation where each agent i pays the cost of the edge leaving agent i on the unique connected path to the root. That is, for every i, we have

$$x_i^B = k_i. \tag{2.17}$$

A weaker version stating that the cost share of agent i should only depend on whether the subtree of downstream agents is empty or not (but not on the structure or the edge costs of the subtree) will lead to a generalization the sequential λ-contributions rule ϕ^λ to standard fixed-tree problems.

Let $\delta(i)$ be the degree of node i (see chapter 1). For $i \in N$, node i is called a *terminal node* if $\delta(i) = 1$. Agents who are not at terminal nodes pay a share $\lambda \in [0, 1]$ of their edge cost k_i as well as an equal share (between successors of the node prior to i) of any remaining debt, and agents at terminal nodes pay their respective residuals. For an agent i, let $\{1, \ldots, i - 1\}$ denote the nodes prior to node i on i's path to the root, ordered in terms of increasing distance to the root. Then we have

$$x_i^\lambda = \lambda \left(k_i + \frac{(1 - \lambda)k_{i-1}}{\delta(i - 1) - 1} + \cdots + \frac{(1 - \lambda)^{i-1}k_1}{\prod_{z=1}^{i-1}(\delta(z) - 1)} \right), \tag{2.18}$$

for nonterminal agents. For terminal agents we have

$$x_i^\lambda = k_i + \frac{(1 - \lambda)k_{i-1}}{\delta(i - 1) - 1} + \cdots + \frac{(1 - \lambda)^{i-1}k_1}{\prod_{z=1}^{i-1}(\delta(z) - 1)}. \tag{2.19}$$

Again, $\lambda = 1$ gives us the sequential full contributions rule (the *Bird allocation* $x_i^B = x_i^1 = k_i$ for all i), and $\lambda = 0$ gives the rule where agents at terminal nodes pay all relevant costs.

To sum up, both the sequential equal-contributions rule and the sequential full-contributions rule can be straightforwardly extended to standard fixed trees with cost shares determined by (2.16) and (2.17), respectively. This also holds for any member of the family of sequential λ-contributions rules given by (2.18) and (2.19).[9]

2.1.2.2   Fixed-tree game

To the extent that it is relevant to consider the fixed-tree cost-allocation problem as a game (N, c), it seems natural to define the cost $c(S)$ of every coalition $S \subseteq N$ as the minimal cost needed to connect all agents in S to the root via a connected subgraph of the fixed tree. Implicitly, we therefore assume that agents in $N \setminus S$ cannot prevent agents in S from using the their nodes to obtain a connection to the root. Formally, for every $S \subseteq N$,

$$c(S) = \sum_{j \in \cup_{i \in S} p^i} k_j, \tag{2.20}$$

which defines the *fixed-tree game*. Example 2.5 provides an illustration.

Example 2.5 (Determining the game and comparing solutions for a fixed tree)   Consider the following fixed tree.

---

9. For this, as well as the obvious interpretation as branch hierarchies, see Hougaard et al. (2017).

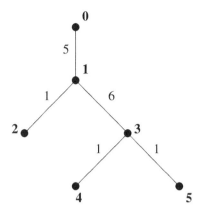

Sharing the total cost of 14 between the five agents using the serial principle (2.16), we get the allocation

$$x^S = (1, 2, 3, 4, 4).$$

Moreover, the fixed-tree game, defined by (2.20), is

$$c(1) = 5, c(2) = 6, c(3) = 11, c(4) = c(5) = 12,$$

$$c(12) = 6, c(13) = 11, c(14) = c(15) = c(23) = 12, c(24) = c(25) = 13,$$

$$c(34) = c(35) = 12, c(45) = 13,$$

$$c(123) = 12, c(124) = 13, c(125) = 13, c(134) = c(135) = 12,$$

$$c(145) = c(234) = c(235) = 13, c(245) = 14, c(345) = 13,$$

$$c(1234) = c(1235) = 13, c(1245) = 14, c(1345) = 13, c(2345) = 14,$$

$$c(12345) = 14.$$

Computing the Shapley value of this game is simple, since i's marginal cost, $c(S \cup i) - c(S)$, is zero for all coalitions S containing a downstream agent of i. Thus we have

$$x^{Sh} = (1, 2, 3, 4, 4),$$

coinciding with the serial allocation above. Note the difference from the Bird allocation, $x^B = (5, 1, 6, 1, 1)$, and the sequential λ-contributions rule for $\lambda = 0.5$,

$$x^{0.5} = (2.5, 2.25, 3.63, 2.81, 2.81).$$

Generally, it can be shown that Shapley value coincides with the sequential equal-contributions rule (2.16), as in the case of the airport problem. The axiomatic characterization in theorems 2.1, 2.2, and 2.3 also extends to the standard fixed-tree framework.[10]

The fixed-tree game, defined by (2.20), is a concave cost-sharing game (like the airport game). A sketch of the argument goes as follows. Let $S \subset T \subseteq N$. For any coalition Z, the subtree connecting $Z \cup i$ to the root contains the subtree connecting Z to the root. Since $S \subset T$, the edges added to the subtree connecting T to the root in order to include i is clearly a subset of the edges added to the subtree connecting S to the root in order to include i. Now, i's marginal cost consists of adding up these extra edge costs, so we get that $c(T \cup i) - c(T) \leq c(S \cup i) - c(S)$, which is tantamount to concavity of c.[11]

Thus, the Shapley value satisfies the core conditions.[12] Clearly, the sequential full-contributions rule (the Bird allocation) also satisfies the core conditions in this case. Koster et al. (2001) provide different characterizations of core elements and consider, in particular, weighted versions of the Lorenz solution. Granot et al. (1996) characterize the nucleolus and provide an algorithm for its computation.

2.1.2.3   Externalities and extended trees

Looking at the airport problem, aside from the fact that larger aircraft types require longer runways, it is typically also the case that they require a higher quality of the runway (broader, thicker, etc). When both small and large aircraft use the same runway, the latter therefore determines the conditions of the runway and, in some sense, forces smaller aircraft to use a runway of higher quality than otherwise needed. As another example, think of users in a multicast data flow requiring different quality of service. In fact, such externalities are present in most practical allocation problems.

Granot et al. (2002) show that it is relatively straightforward to extend the standard fixed-tree model to such a situation with agents of different types $l = 1, \ldots, p$ $(p \leq n)$ in the sense of requiring, for example, different levels of capacity (or quality) of their connection to the root. For each agent $i \in N$, it is required that each edge in the (unique) path from i to the root 0 is capable of satisfying i's type. Denote by γ_i the type of agent i, and assume that the types are ordered by increasing level of capacity (or quality). Moreover, denote by e_i the edge leaving node i in the direction of the root.

---

10. See, e.g., Hougaard et al. (2017).

11. In the case where the agents can be located at multiple nodes, the (modified) fixed-tree problem need no longer be concave (Miquel et al. 2006).

12. In fact, for a concave cost game, the core equals the set of weighted Shapley values; see, e.g., Monderer et al. (1992) or Bjørndal et al. (2004).

Given agent $i \in N$, let $\mathcal{E}_i \subseteq N$ denote the set of agents for whom the unique path to the root includes the edge e_i, and for every $S \subseteq N$, let $\gamma(S) = \max\{\gamma_i \mid i \in S\}$ be the highest indexed type in coalition S. Thus, the required capacity (or quality) of e_i in the optimal subtree for coalition $S \subseteq N$ must be $\gamma(\mathcal{E}_i \cap S)$, since the capacity of edge e_i must be able to satisfy the most demanding downstream agent in coalition S.

Let a_i^l be the cost of using edge e_i when i is of type l, and assume that costs are increasingly ordered by type, that is, $0 = a_i^0 \leq a_i^1 \leq \cdots \leq a_i^p$ for each agent $i \in N$. Thus, the associated cost-allocation game can be defined as $(N, \hat{c})$, where

$$\hat{c}(S) = \sum_{i \in N} a_i^{\gamma(\mathcal{E}_i \cap S)} \tag{2.21}$$

for all $S \subseteq N$. Consider the following example.

Example 2.5 (continued: Including externalities)   Consider again the tree

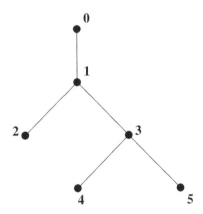

Assume there are four types, where agent 1 is of type 2, agent 2 is of type 4, agent 3 is of type 1, agent 4 is of type 2, and agent 5 is of type 3. We therefore have

$\hat{c}(1) = a_1^2,$

$\hat{c}(2) = a_1^4 + a_2^4,$

$\hat{c}(3) = a_1^1 + a_3^1,$

$\hat{c}(4) = a_1^2 + a_3^2 + a_4^2,$

$\hat{c}(5) = a_1^3 + a_3^3 + a_5^3,$

$$\hat{c}(12) = a_1^4 + a_2^4,$$

$$\hat{c}(13) = a_1^2 + a_3^1,$$

$$\hat{c}(14) = a_1^2 + a_3^2 + a_4^2,$$

$$\hat{c}(15) = a_1^3 + a_3^3 + a_5^3,$$

$$\hat{c}(23) = a_1^4 + a_2^4 + a_3^1,$$

$$\hat{c}(24) = a_1^4 + a_2^4 + a_3^2 + a_4^2,$$

$$\hat{c}(25) = a_1^4 + a_2^4 + a_3^3 + a_5^3,$$

$$\hat{c}(34) = a_1^2 + a_3^2 + a_4^2,$$

$$\hat{c}(35) = a_1^3 + a_3^3 + a_5^3,$$

$$\hat{c}(45) = a_1^3 + a_3^3 + a_4^2 + a_5^3,$$

$$\hat{c}(123) = a_1^4 + a_2^4 + a_3^1,$$

$$\hat{c}(124) = a_1^4 + a_2^4 + a_3^2 + a_4^2,$$

$$\hat{c}(125) = a_1^4 + a_2^4 + a_3^3 + a_5^3,$$

$$\hat{c}(134) = a_1^2 + a_3^2 + a_4^2,$$

$$\hat{c}(135) = a_1^3 + a_3^3 + a_5^3,$$

$$\hat{c}(145) = a_1^3 + a_3^3 + a_4^2 + a_5^3,$$

$$\hat{c}(234) = a_1^4 + a_2^4 + a_3^2 + a_4^2,$$

$$\hat{c}(235) = a_1^4 + a_2^4 + a_3^3 + a_5^3,$$

$$\hat{c}(245) = a_1^4 + a_2^4 + a_3^3 + a_4^2 + a_5^3,$$

$$\hat{c}(345) = a_1^3 + a_3^3 + a_4^2 + a_5^3,$$

$$\hat{c}(1234) = a_1^4 + a_2^4 + a_3^2 + a_4^2,$$

$$\hat{c}(1235) = a_1^4 + a_2^4 + a_3^3 + a_5^3,$$

$$\hat{c}(1245) = a_1^4 + a_2^4 + a_3^3 + a_4^2 + a_5^3,$$

$$\hat{c}(1345) = a_1^3 + a_3^3 + a_4^2 + a_5^3,$$

$$\hat{c}(2345) = a_1^4 + a_2^4 + a_3^3 + a_4^2 + a_5^3,$$

$$\hat{c}(12345) = a_1^4 + a_2^4 + a_3^3 + a_4^2 + a_5^3.$$

It is easy to show that a naive extension of the Bird allocation to the present framework will no longer result in core allocations. If the total cost $\hat{c}(12345) = a_1^4 + a_2^4 + a_3^3 + a_4^2 + a_5^3$ is allocated by letting agent i pay for the edge e_i, so that $(x_1, x_2, x_3, x_4, x_5) = (a_1^4, a_2^4, a_3^3, a_4^2, a_5^3)$, then we get, for instance, that $x_1 = a_1^4 > \hat{c}(1) = a_1^2$, violating the core conditions (and thereby the standalone principle).

If the cost graph is a chain with $p = n$ (as in the airport game), then the standalone cost of each type $i \in N$ is $C_i = \sum_{j \leq i} a_j^i$, so the Shapley value of $(N, \hat{c})$ is determined as: $\phi_1^{Sh}(N, \hat{c}) = \frac{1}{n} a_1^1$, $\phi_2^{Sh}(N, \hat{c}) = \frac{1}{n} a_1^1 + \frac{1}{n-1}(a_2^2 + a_1^2 - a_1^1)$, and so on, in accordance with the serial principle. In comparison, the naive Bird allocation is given by $x_i = a_i^n$ for every $i \in N$.

By a similar type of argument as in the case of standard fixed trees, it can be shown that the cost-sharing game $(N, \hat{c})$ associated with the extended tree is concave. Hence, the core is nonempty and contains the Shapley value, also when the cost-sharing game accounts for externalities. Granot et al. (2002) further characterize core elements and provide an algorithm for checking whether a given allocation satisfies the core conditions.

2.2 Minimum-Cost Spanning Trees

Now consider a more general scenario involving the complete graph from the outset. A group of agents, each located at a different node, want to be connected to a common supplier (the root). Each undirected connection—between agents or between an agent and the supplier—has a constant cost (i.e., a cost that is independent of the number of agents using the connection). Then connections (edges) are public goods, and we may think of the costs as the cost of establishing the connection or as maintenance costs.

Assuming there are no obvious externalities involved, the two main problems are

1. To find the cost-minimizing network connecting all agents to the supplier (either directly or indirectly via other agents); and

2. To allocate the total cost of the resulting (cost-minimal) network among the agents.

Thus, there is no fixed network from the start: the problem is one of optimal network design, finding a minimum-cost (efficient) graph. Even though it is clear that the resulting efficient graph becomes a tree, the fact that we now know the cost of all edges in the complete graph radically changes the cost-allocation problem compared to the fixed-tree situation analyzed above. This will become clear as we proceed.

Practical examples include district heating, where a power plant (the supplier) delivers heating to a set of households (the agents) via a network of insulated pipes; cable TV, where a broadcaster (the supplier) transmits television programs to a set of consumers (the agents) via a network of coaxial cables; computer networks (the agents) using a common server (the supplier); and chain stores (the agents) using a common warehouse (the supplier).

2.2.1 The Model

Let $N = \{1, \dots, n\}$ denote the set of agents and 0 denote the supplier (root). Moreover, let $N^0 = N \cup \{0\}$ and G^{N^0} be the complete rooted graph. For each edge $ij \in G^{N^0}$, let $k_{ij} > 0$ be a positive constant cost associated with edge ij. Since edges are undirected, $k_{ij} = k_{ji}$. Setting $k_{ii} = 0$, for all $i \in N^0$, define a $(n+1) \times (n+1)$ cost matrix K associated with the complete rooted graph G^{N^0}. We write $K \leq K'$ if $k_{ij} \leq k'_{ij}$ for all $i, j \in N^0$. The pair (N, K) is dubbed a *minimum-cost spanning-tree problem* (in short, a MCST problem).

Recall the relevant notions and results from Chapter 1. A *spanning tree* $T \in \mathcal{T}_{N^0}$ is a tree where all n agents are connected to the root, 0, either directly or indirectly. There are $(n+1)^{n-1}$ spanning trees, each with n edges. For a given MCST problem (N, K), a *minimum-cost spanning tree* is a spanning tree whose total cost $\sum_{ij \in T} k_{ij}$ is minimized over all possible spanning trees:

$$
\mathcal{T}^{min}(N, K) = \left\{ \arg\min_{T \in \mathcal{T}_{N^0}} \sum_{ij \in T} k_{ij} \right\}. \tag{2.22}
$$

In other words, a MCST $T \in \mathcal{T}^{min}$ is an efficient network connecting all agents to the root. There is at least one, but there may be multiple MCSTs for a given MCST problem (indeed, if all link costs are identical, all spanning trees are minimum cost). As mentioned in chapter 1, a sufficient condition for the existence of a unique MCST is that no two edge costs are the same.

For a given MCST, $T \in \mathcal{T}^{min}$, there is a unique path connecting a given agent i to the root (since a tree contains no cycles) with exiting edge e_i.

Finding MCSTs is simple, using, for example, the Prim or the Kruskal algorithm (see chapter 1).

2.2.2 Allocation in Spanning Trees

Once the set of MCSTs has been identified, it remains to allocate the total
network cost among the agents. This issue was initially discussed in Claus and
Kleitman (1973) and further analyzed in Bird (1976).

For a given MCST $T \in \mathcal{T}^{min}$, the problem of allocating the total cost resembles
that of cost allocation for standard fixed trees. Thus, the serial allocation (called
"allocation by actual cost" in Claus and Kleitman 1973) defined by (2.16), the
Bird allocation defined by (2.17), and indeed any sequential λ-contributions rule
defined by (5.11) and (2.19), can be directly applied on MCSTs, too.

For convenience, denote by $x(T)$ the (budget-balanced) cost allocation associated with the MCST $T \in \mathcal{T}^{min}$. Since there may be more than one MCST for
a given MCST problem (N, K), these allocations do not constitute allocation
rules (in the sense that they do not present a unique solution for every MCST
problem).

Yet, obviously the serial allocation and the Bird allocation (or any other allocation defined for a specific spanning tree) can be turned into allocation rules.
For instance, we can take a simple average over all the resulting allocations
for a given MCST problem (or any other convex combination of tree-specific
allocations). That is, we can define the serial rule, ϕ^S, and Bird rule, ϕ^B,
respectively, as follows:

$$\phi^S(N, K) = \frac{1}{|\mathcal{T}^{min}|} \sum_{T \in \mathcal{T}^{min}} x^S(T), \tag{2.23}$$

where $x^S(T)$ is determined by (2.16), and

$$\phi^B(N, K) = \frac{1}{|\mathcal{T}^{min}|} \sum_{T \in \mathcal{T}^{min}} x^B(T), \tag{2.24}$$

where $x^B(T)$ is determined by (2.17).

Example 2.6 (Finding MCSTs and computing standard solutions) Consider the following complete cost graph for agents $N = \{a, b, c\}$.

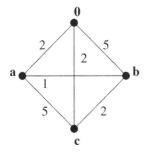

Clearly, there are three MCSTs $T = \{0a, ab, bc\}$, $T' = \{0c, cb, ba\}$, and $T'' = \{0a, 0c, ab\}$, each with a total (minimal) connection cost of 5, as shown below.

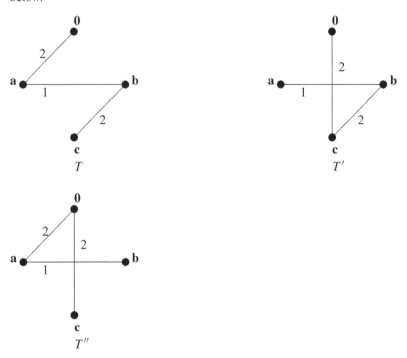

Using the Bird allocation with respect to T and T'', we get cost shares $x^B(T) = x^B(T'') = (2, 1, 2)$, while the Bird allocation with respect to T' is given by $x^B(T') = (1, 2, 2)$. Consequently, the Bird *rule* (2.24) results in cost shares

$$\phi^B(N, K) = (1.67, 1.33, 2).$$

Using the serial principle results in the following cost shares for the (chain) T:

$$x_a^S(T) = (1/3)k_{0a} = 0.67,$$

$$x_b^S(T) = (1/3)k_{0a} + (1/2)k_{ab} = 0.67 + 0.5 = 1.17,$$

$$x_c^S(T) = (1/3)k_{0a} + (1/2)k_{ab} + k_{bc} = 0.67 + 0.5 + 2 = 3.17.$$

Likewise, considering (the chain) T' we have

$$x_a^S(T') = (1/3)k_{0c} + (1/2)k_{cb} + k_{ba} = 0.67 + 1 + 1 = 2.67,$$

$$x_b^S(T') = (1/3)k_{0c} + (1/2)k_{cb} = 0.67 + 1 = 1.67,$$

$$x_c^S(T') = (1/3)k_{0c} = 0.67.$$

Finally, considering (the tree) T'' we have

$$x_a^S(T'') = (1/2)k_{0a} = 1,$$

$$x_b^S(T'') = (1/2)k_{0a} + k_{ab} = 1 + 1 = 2,$$

$$x_c^S(T'') = k_{0c} = 2.$$

Note that when using the serial allocation for T, agent c will pay more than his standalone cost $k_{0c} = 2$. Likewise, considering T', agent a pays more than her standalone cost $k_{0a} = 2$. This is in contrast to the standard fixed-tree scenario, where use of the sequential equal-contributions rule will lead to core allocations.

Consequently, the serial *rule* (2.23) results in cost shares

$$\phi^S(N, K) = (1.44, 1.61, 1.94).$$

Allocations (and thereby allocation rules) may also be derived from the algorithms for finding MCSTs. For instance, the Bird allocation can be seen as derived from the Prim algorithm. Recall example 2.6 above.

Example 2.6 (continued: Using the Prim algorithm to determine a Bird allocation) Start out with the root 0. Following the Prim algorithm, we select the cheapest connection among agents $\{a, b, c\}$ to the root 0. In this case, a and c both have edge cost $k_{0a} = k_{0c} = 2$. Allocating each agent their respective edge costs results in two allocations: one where $x_a = 2$, and one where $x_c = 2$. Consider the former. The Prim algorithm states that we select the cheapest link between an agent in $\{b, c\}$ to either 0 or a: here it is the link ab with $k_{ab} = 1$. Hence, b will be allocated 1, and the final allocation $(2, 1, 2)$ corresponds to the Bird allocation for T (the resulting MCST). Now consider the other possibility, where $x_c = 2$. Following the Prim algorithm, we select the cheapest link among $\{a, b\}$ to either 0 or c: that is, either connect a to 0, or b to c (both with cost 2). Hence, two allocations prevail: the first gives $(2, 1, 2)$ (i.e., the Bird allocation of the resulting MCST T''); the second gives $(1, 2, 2)$ (i.e., the Bird allocation of the resulting MCST T').

The Prim algorithm can also lead to other allocations, as suggested in Dutta and Kar (2004). For simplicity, assume that all link costs differ. (The algorithm below easily generalizes to arbitrary MCST problems using some ordering of N as a tie-breaking rule in case several links all minimize the cost of connecting nodes in A^{k-1} and A_c^{k-1} of step k. Then a rule can be defined averaging over the allocations obtained for the different orderings.) On this domain there is a unique MCST, and the Prim algorithm can be used to define the following allocation rule ϕ^{DK}:

Let $A^0 = \{0\}$, $g^0 = \emptyset$ and $t^0 = 0$.

Step 1. Choose the pair $a^1 b^1 = \operatorname{argmin}_{ij \in A^0 \times A_c^0} k_{ij}$, where $A_c = N^0 \setminus A$. Define $t^1 = \max\{t^0, k_{a^1 b^1}\}$, $A^1 = A^0 \cup \{b^1\}$, and $g^1 = g^0 \cup \{a^1, b^1\}$.

Step h. Choose the pair $a^h b^h = \operatorname{argmin}_{ij \in A^{h-1} \times A_c^{h-1}} k_{ij}$, and define $t^h = \max\{t^{h-1}, k_{a^h b^h}\}$, $A^h = A^{h-1} \cup \{b^h\}$, and $g^h = g^{h-1} \cup \{a^h, b^h\}$.

Now define

$$\phi_{b^{h-1}}^{DK}(N, K) = \min\{t^{h-1}, k_{a^h b^h}\}. \tag{2.25}$$

The algorithm stops at step n with

$$\phi_{b^n}^{DK}(N, K) = t^n. \tag{2.26}$$

Thus, according to the rule ϕ^{DK}, the agent (b^{k-1}) being connected in step $k - 1$ pays the minimum of t^{k-1} (which is the maximum cost among the links already constructed) and the cost $(k_{a^k b^k})$ of the link being constructed in step k. The last agent to be connected pays the maximum link cost t^n. At first glance, this seems a somewhat strange way of defining an allocation rule, but the reason will become clear in section 2.2.4. To illustrate the idea, consider the following example.

Example 2.7 (Illustrating the Dutta-Kar solution) Consider a case with three agents, $N = \{a, b, c\}$, where the MCST is a chain, for example, $T^{min} = \{0a, ab, bc\}$. Using the Prim algorithm, we have

Step 1. $a^1 b^1 = 0a$, making $t^1 = k_{0a}$, $A^1 = \{0, a\}$, and $g^1 = \{0a\}$.

Step 2. $a^2 b^2 = ab$, making $t^2 = \max\{k_{0a}, k_{ab}\}$, $A^2 = \{0, a, b\}$, and $g^2 = \{0a, ab\}$.

Step 3. $a^3 b^3 = bc$, making $t^3 = \max\{k_{0a}, k_{ab}, k_{bc}\}$, $A^3 = \{0, a, b, c\}$, and $g^3 = \{0a, ab, bc\}$.

Thus, we have

$$\phi_a^{DK}(N, K) = \min\{k_{0a}, k_{ab}\},$$

$$\phi_b^{DK}(N, K) = \min\{\max\{k_{0a}, k_{ab}\}, k_{bc}\},$$

$$\phi_c^{DK}(N, K) = \max\{k_{0a}, k_{ab}, k_{bc}\}.$$

This rule clearly differs from the Bird allocation (e.g., when $k_{ab} < k_{0a}$).

Allocations can also be derived from the Kruskal algorithm (e.g., as analyzed in Tijs et al. 2006, Lorenzo and Lorenzo-Freire 2009, and Bergantiños et al. 2010). Consider example 2.6 again.

Example 2.6 (continued: Using the Kruskal algorithm to determine an allocation) Following the Kruskal algorithm, the first edge to be formed is ab with a cost of 1. Continuing, there are several options, since the next edges to be added (either $\{0a, bc\}$, $\{bc, 0c\}$ or $\{0a, 0c\}$) all have a cost of 2.

The idea is that agents pay for each additional edge according to their remaining *obligations*. The obligation of a given agent in a coalition $S \subset N^0$ is $1/|S|$ if $0 \notin S$, and zero otherwise. By adding edges one by one, agents either enter or remain in the coalition of connected agents. For every edge, all connected agents therefore have a remaining obligation (the difference between the obligation before and after adding the edge).

Consider the first case (corresponding to the MCST T in the original version of example 2.6): Before adding the edge ab, all agents have obligation 1. After adding the edge, a and b become connected, so their remaining obligation is $(1 - 0.5 =)$ 0.5. Hence, the cost $k_{ab} = 1$ is shared equally between a and b. Adding the edge $0a$ introduces a new cost $k_{0a} = 2$. The remaining obligation for both a and b is $(0.5 - 0 =)$ 0.5, so this cost also is shared equally between a and b. Finally, adding the edge bc gives c (indirect) access to the root, while both a and b are indifferent to the link. Remaining obligations for the three agents are therefore $(0, 0, 1)$. Hence, c has to cover the entire cost alone. The resulting allocation thus becomes $(1.5, 1.5, 2)$.

Consider the second case (corresponding to the MCST T'): again a and b have to share the costs of the first edge ab. Adding bc to ab gives the remaining obligations $(\frac{1}{6}, \frac{1}{6}, \frac{2}{3})$, so agent a pays $\frac{1}{6}$, agent b pays $\frac{1}{6}$, and agent c pays $\frac{4}{3}$ for the edge bc. Finally, adding $0c$ gives the remaining obligations $(\frac{1}{3}, \frac{1}{3}, \frac{1}{3})$, so the cost $k_{0c} = 2$ is shared equally among the three agents. The resulting allocation is again $(1.5, 1.5, 2)$.

Finally, consider the third case (corresponding to the MCST T'' in the original version of example 2.6): Again a and b share equally the cost of the first edge ab. Adding $0a$ to ab gives remaining obligations 0.5 for both agents a and b. Hence, they also share the cost $k_{0a} = 2$ equally. Finally, adding $0c$ only benefits c who has remaining obligation 1 and hence carries the cost alone. The resulting allocation again becomes $(1.5, 1.5, 2)$. The fact that all three trees result in the same allocation is not a coincidence: the resulting allocation is known as the Equal Remaining Obligations rule, or the P-value, in the literature (Feltkamp et al., 1994, Branzei et al., 2004). Below, a broader class of so-called *obligation* rules will be defined.

The example above demonstrates how rules can be constructed in the spirit of the Kruskal algorithm, sharing the costs of the added links one by one according to equal remaining obligations. This approach can be generalized as shown in Bergantiños et al. (2010). Formally, define the class of *obligation rules*, ϕ^o, as follows.

Consider a coalition $S \subset N^0$, and define the obligation o by the following mapping. If $0 \in S$, the obligation is $o_i(S) = 0$ for all $i \in S \cap N$. If $0 \notin S$, the obligation is $o(S) \in \{x \in \mathbb{R}_+^S \mid \sum_{i \in S} x_i = 1\}$, and for each $S \subset T$ and each $i \in S$, $o_i(S) \le o_i(T)$.

Moreover, let $(lj)^p$ be the pth edge added to the graph g^{p-1} of connected components resulting from the Kruskal sequence of edges up to step $p - 1$. Let $S(g)$ be the coalition of agents $S \subset N^0$ that are connected in graph g. Then the class of obligation rules is given by allocations

$$x_i^o = \sum_{p=1}^{n} k_{(lj)^p} (o_i(S(g^{p-1})) - o_i(S(g^p))). \tag{2.27}$$

Consider example 2.6 again.

Example 2.6 (continued: Using (2.27)) Following the general formula (2.27), we get the allocation

$$\phi^0(N, K) = (1 + o_a(ab), 1 + o_b(ab), 2),$$

where $o_a(ab) + o_b(ab) = 1$. In particular, for the specific tree T (in the original example 2.6) we have

$$x_a^o(T) = 1(o_a(a) - o_a(ab)) + 2(o_a(ab) - o_a(0ab)) + 2(o_a(0ab) - o_a(0abc))$$
$$= 1 + o_a(ab),$$
$$x_b^o(T) = 1(o_b(b) - o_b(ab)) + 2(o_b(ab) - o_b(0ab)) + 2(o_b(0ab) - o_b(0abc))$$
$$= 1 + o_b(ab),$$
$$x_c^o(T) = 2(o_c(c) - o_c(0abc)) = 2.$$

As mentioned above, the allocation is the same regardless of the chosen MCST. For instance, for the tree T', we have

$$x_a^o(T) = 1(o_a(a) - o_a(ab)) + 2(o_a(ab) - o_a(abc)) + 2(o_a(abc) - o_a(0abc))$$
$$= 1 + o_a(ab),$$
$$x_b^o(T) = 1(o_b(a) - o_b(ab)) + 2(o_b(ab) - o_b(abc)) + 2(o_b(abc) - o_b(0abc))$$

$$= 1 + o_b(ab),$$
$$x_c^o(T) = 2(o_c(c) - o_c(abc)) + 2(o_c(abc) - o_c(0abc)) = 2.$$

The class of obligation rules turns out to satisfy some very strong mono-tonicity properties, as we will see in section 2.2.4.

2.2.3 The MCST Problem as a Game

The MCST problem can also be modeled by a cooperative game (as originally suggested in Bird, 1976). In many ways this seems natural, since in a design phase, the efficient network can only be realized if all involved agents accept their assigned cost shares, which in turn depends on their outside options (establishing local networks). So the challenge becomes how to determine the agents' outside options.

For a given MCST problem, there are at least two relevant ways of deter-mining these options (Granot and Huberman, 1981): one is to consider each coalition of agents $S \subseteq N$ in isolation and determine the minimum cost of con-necting coalition S to the root (using the projection of the cost matrix K to S^0); the other is to allow agents in coalition S to connect to the root via agents outside the coalition (i.e., among agents in the complement $N \setminus S$).

In the former case, the MCST problem (N, K) can be represented as a cost-allocation problem (N, c), where N is the set of agents, and c is a cost function defined by

$$c(S) = \min_{T \in \mathcal{T}_{S^0}} \sum_{ij \in T} k_{ij}, \tag{2.28}$$

for all $S \subseteq N$. In the latter case, the MCST problem (N, K) can be represented as a cost-allocation problem $(N, \bar{c})$, where the cost function $\bar{c}$ is defined by

$$\bar{c}(S) = \min_{T \supseteq S} c(T), \tag{2.29}$$

for all $S, T \subseteq N$.

Even though there may be several MCSTs related to a given subgraph, the cost functions c and $\bar{c}$ are uniquely determined for a given MCST problem. Moreover, it is clear that $c(N) = \bar{c}(N)$ and $c(S) \geq \bar{c}(S)$ for all coalitions S.

Example 2.6 (continued: Two interpretations) Modeling the MCST pro-blem in example 2.6 as a cooperative game, we obtain the cost-allocation problems (N, c) and $(N, \bar{c})$, respectively:

$$c(a) = 2, \quad c(b) = 5, \quad c(c) = 2,$$
$$c(ab) = 3, \quad c(ac) = 4, \quad c(bc) = 4,$$

$c(abc) = 5.$

$\bar{c}(a) = 2, \quad \bar{c}(b) = 3, \quad \bar{c}(c) = 2,$

$\bar{c}(ab) = 3, \quad \bar{c}(ac) = 4, \quad \bar{c}(bc) = 4,$

$\bar{c}(abc) = 5.$

The two-agent subgraphs are given as below with total minimal cost of 3, 4, and 4, respectively.

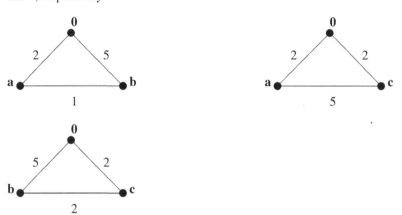

The only difference between (N, c) and $(N, \bar{c})$ is that agent b in the latter case is allowed to connect to the root via agent a, and hence is able to reduce the standalone cost from 5 to 3. No other coalition can gain by connecting via outside agents in this case.

Contrary to the standard fixed-tree model, neither of the cost-allocation problems (N, c) and $(N, \bar{c})$ are concave. Hence, for instance, the Shapley value is not guaranteed to satisfy the core conditions for all problems.

Yet for every MCST problem, any Bird allocation x^B, and thereby also the Bird rule ϕ^B, satisfies the core conditions of both derived games (N, c) and $(N, \bar{c})$ (Bird, 1976; Granot and Huberman, 1981).

Theorem 2.4 Both (N, c) and $(N, \bar{c})$ are balanced problems (i.e., the core is nonempty). Moreover, $core(N, \bar{c}) \subseteq core(N, c)$.

Sketch of proof: Focus on the game (N, c) and let x^B be any Bird allocation of the original MCST problem. Suppose that x^B is based on a MCST with e_i^B being the leaving edge for every agent i. Consider an arbitrary coalition $S \subseteq N$. Construct a graph $T = T^{min}(S, K_{|S}) \cup \{e_i^B\}_{i \in N \setminus S}$. Then T is a spanning tree. Thus, we have

$$\sum_{ij \in T} k_{ij} = c(S) + \sum_{i \in N \setminus S} x_i^B \geq \sum_{ij \in T^{min}(N,K)} k_{ij} = c(N)$$

$$\Rightarrow c(S) \geq \sum_{i \in S} x_i^B.$$

Given that $\bar{c}$ is balanced, $core(N, \bar{c}) \subseteq core(N, c)$, since $\bar{c}(S) \leq c(S)$ for all $S \subseteq N$.

Example 2.6 (continued: Simple solutions based on standalone costs)
Since, for every MCST problem, $\sum_{i \in N} c(i) \geq c(N)$, we can use simple rationing rules (see, e.g., Hougaard, 2009; Thomson, 2015) to allocate the total cost $c(N)$. Looking at the game (N, c) from example 2.6, we can, for instance, share the total cost of 5 equally, provided no one pays more than their stand-alone cost (a rule called the constrained equal awards rule in rationing), or dually, share the excess cost of $2 + 5 + 2 - 5 = 4$ equally, provided no agent is subsidized (a rule called the constrained equal loss rule in rationing). We then get, respectively,

$$x^{CEA} = (1.66, 1.66, 1.66), \quad \text{and} \quad x^{CEL} = (0.66, 3.66, 0.66).$$

But we can also choose to share in proportion to standalone costs, which results in the payments

$$x^P = (1.11, 2.78, 1.11).$$

None of these solutions are core allocations. In this case, the core conditions of the game (N, c) dictate that $x_a + x_b = 3$, and $1 \leq x_a \leq 2$, $1 \leq x_b \leq 5$, $x_c = 2$. These condition are violated by all three allocations above. Since the core is nonempty, the relevance of these solutions is therefore questionable.

The cost function $\bar{c}$ is monotone; that is, if $S \subseteq T$, then $\bar{c}(S) \leq \bar{c}(T)$. Thus, $core(N, \bar{c}) \in \mathbb{R}_+^n$. In general, this is not the case for c.

Example 2.8 (The games c and $\bar{c}$ are not concave, and c is not monotone)
Consider the following complete weighted-cost graph for agents $N = \{a, b, c\}$.

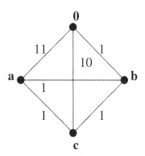

Thus, the respective games become

$$c(a) = 11, \qquad \bar{c}(a) = 2,$$
$$c(b) = 1, \qquad \bar{c}(b) = 1,$$
$$c(c) = 10, \qquad \bar{c}(c) = 2,$$
$$c(ab) = 2, \qquad \bar{c}(ab) = 2,$$
$$c(ac) = 11, \qquad \bar{c}(ac) = 3,$$
$$c(bc) = 2, \qquad \bar{c}(bc) = 2,$$
$$c(abc) = 3, \qquad \bar{c}(abc) = 3.$$

Clearly, c is neither concave ($c(abc) + c(a) = 14 > 13 = c(ab) + c(ac)$) nor monotone (e.g., $c(ab) = 2 < 11 = c(a)$), and we may have core allocations with negative values (e.g., $(5.5, -8, 5.5) \in core(N, c)$). On the contrary, in this case $\bar{c}$ is concave and monotone (by definition), with all core allocations having positive values. Note that letting, for example, $k_{0c} = 1$ instead of 10 (*ceteris paribus*) results in games (N, c) and $(N, \bar{c})$, where neither game is concave.

Recall from chapter 1 that there is at least one more interesting game that can be derived from a MCST problem (N, K): let c^* be the cost function derived from the irreducible cost matrix K^*, that is, the smallest matrix for which (any) MCST $T \in \mathcal{T}^{min}$ remains a MCST (Bird, 1976). That is, for all $S \subseteq N$,

$$c^*(S) = \min_{T \in \mathcal{T}_{S^0}} \sum_{ij \in T} k_{ij}^*. \tag{2.30}$$

Recall that $k_{ij}^* = \max_{lm \in T_{ij}^{min}} k_{lm}$, where T_{ij}^{min} is the unique path in $T \in \mathcal{T}^{min}$ connecting i and j. It was noted in chapter 1 that K^* (and therefore c^*) is uniquely determined for a given MCST problem (N, K). Let (N, c^*) be the game representing the irreducible problem (N, K^*). Clearly, we have $c^*(S) \leq \bar{c}(S) \leq c(S)$ for all $S \subset N$, and $c^*(N) = \bar{c}(N) = c(N)$ (indeed, $c^*(S) \leq \bar{c}(S)$, since no edge cost k_{ij}^* for $ij \in S^0$ can be replaced by a cheaper route connecting i and j via agents in $N \setminus S$. Therefore, it also follows that c^* is monotone).

Contrary to both c and $\bar{c}$, the (irreducible) cost function c^* is concave (indeed, we can show that $c^*(S \cup i) - c^*(S) = \min_{j \in S^0} k_{ij}^*$, and since $\min_{j \in T^0} k_{ij}^* \leq \min_{j \in S^0} k_{ij}^*$ for $S \subset T$, concavity of c^* follows).

Thus, $core(N, c^*) \subseteq core(N, \bar{c}) \subseteq core(N, c)$. In fact, it can be shown (Bird, 1976) that the core of (N, c^*) is identical to the convex hull of all Bird allocations for (N, K):

$$core(N, c^*) = conv\{x^B(T)\}_{T \in \mathcal{T}^{min}(N,K)}.$$ (2.31)

Indeed, since c^* is concave, $core(N, c^*) = web(N, c^*)$ (see chapter 1). For $i \in N$, we have $x_i^\pi = \min_{j \in \{S_{\pi,i-1} \cup 0\}} k_{ij}^*$. Hence, the allocations x^π are Bird allocations.

Example 2.6 (continued: The core of c^*)   Recall the complete cost graph of example 2.6.

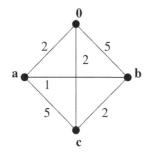

The corresponding irreducible costs are shown in the following graph.

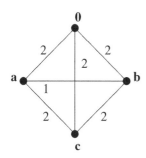

Therefore, we have the cost-allocation problem (N, c^*), where c^* is given by

$$c^*(a) = c^*(b) = c^*(c) = 2,$$
$$c^*(ab) = 3, \ c^*(ac) = c^*(bc) = 4,$$
$$c^*(abc) = 5.$$

The core conditions are

$$1 \le x_a \le 2,$$
$$1 \le x_b \le 2,$$
$$2 \le x_c \le 2, \quad \text{where } x_a + x_b + x_c = 5.$$

Hence,

$$core(N, c^*) = \lambda(1, 2, 2) + (1 - \lambda)(2, 1, 2), \quad \lambda \in [0, 1],$$

that is, the convex hull of the two distinct Bird allocations.

Since c^* is concave, the Shapley value of the irreducible game (N, c^*) will be a core solution for all cost functions, c^*, $\bar{c}$, and c. Thus, we can use the Shapley value of (N, c^*) as a solution to the MCST problem (N, K): the so-called *Folk solution*:

$$\phi^{Folk}(N, K) = \phi^{Sh}(N, c^*). \tag{2.32}$$

It turns out that the equal remaining obligations rule coincides with the Folk solution. Moreover, $\phi^{Folk}(N, K) = \phi^B(N, K^*)$. In fact, it is particularly simple to compute the Shapley value of (N, c^*) (and thereby the Folk solution), since we can write

$$\phi_i^{Sh}(N, c^*) = \frac{1}{n} k_{0i}^* + \sum_{j=1}^{n-1} \frac{1}{j(j+1)} \min\{k_i^{*j}, k_{0i}^*\}, \tag{2.33}$$

where k_i^{*j} is agent i's irreducible connection cost to the $n - 1$ other agents ordered by increasing value, that is, $k_i^{*1} \leq \cdots \leq k_i^{*n-1}$ (Bergantiños and Vidal-Puga, 2007; Bogomolnaia and Moulin, 2010).

2.2.4   Axioms and Characterizations

When making a qualitative assessment of alternative allocation rules, one important issue concerns their monotonicity properties. That is, how cost shares are influenced by varying connection costs and population size. We start out by considering two standard versions of cost monotonicity and a standard version of population monotonicity.

Cost monotonicity   For all pairs of cost matrices K and K', where $k_{ij} > k'_{ij}$ for $i, j \in N^0$ and $k'_{nm} = k_{nm}$ for all $\{n, m\} \neq \{i, j\}$,

$$\phi_z(N, K) \geq \phi_z(N, K'),$$

for each $z \in N \cap \{i, j\}$.

Cost monotonicity states that when the edge cost between any two nodes in N^0 increases, the cost share of the agents involved weakly increases. Note that one of the nodes may be the (nonpaying) root 0.

Strong cost monotonicity For all pairs of cost matrices K and K' where $K \geq K'$,

$$\phi(N, K) \geq \phi(N, K').$$

Strong cost monotonicity states that no agent can gain when connection costs increase. Clearly, strong cost monotonicity implies cost monotonicity. It turns out that cost monotonicity (and strong cost monotonicity) plays a crucial part in the implementation of efficient networks (MCSTs), as will become clear in Chapter 6.[13]

Population monotonicity For all MCST problems (N, K), $S \subset N$, and $z \in S$,

$$\phi_z(N, K) \leq \phi_z(S, K_{|S}),$$

where $K_{|S}$ is the projection of K onto S^0.

Population monotonicity implies that when society expands, none of the original members will be worse off.

Example 2.9 (Violations of (strong) cost monotonicity) Consider two problems involving agents Ann and Bob ($N = \{a, b\}$) with $K' = (3, 2, 1)$ and $K = (3, 4, 1)$, as illustrated below. Note that the problems only differ in Bob's connection cost to the root (being $k'_{0b} = 2$, and $k_{0b} = 4$).

In (N, K'), the unique MCST is given by $T = \{0b, ba\}$ with total cost 3: hence the unique Bird allocation is $x^B(T) = (1, 2)$.

In (N, K), the unique MCST is given by $T' = \{0a, ab\}$ with total cost 4: hence the unique Bird allocation is $x^B(T') = (3, 1)$.

So even though the link cost increases for Bob ($k'_{0b} = 2 < 4 = k_{0b}$) when going from K' to K, he now pays less. Consequently, the Bird rule violates cost monotonicity.

Using the allocation rule ϕ^{DK} (given by (2.25) and (2.26)), we get $\phi^{DK}(N, K') = (2, 1)$ and $\phi^{DK}(N, K) = (1, 3)$. So in this case, Bob's payment

---

13. See Hougaard and Tvede (2012).

increases as a result of his increasing cost of connecting to the source. In fact, Dutta and Kar (2004) show that ϕ^{DK} satisfies cost monotonicity in general. Like the Bird rule, ϕ^{DK} produces an allocation satisfying the core conditions. Yet this example makes it clear that ϕ^{DK} violates strong cost monotonicity, since Ann is better off by the increase in Bob's connection cost to the root.

Allocating costs according to the serial rule (using (2.16)) also violates cost monotonicity. Consider a network with five agents $N = \{a, b, c, d, e\}$ and connection costs given by $k_{0a} = k_{ab} = k_{0c} = 2$, $k_{cd} = k_{de} = 0$, $k_{eb} = 2.1$, and $k_{..} = \infty$ otherwise. Consequently there is a unique MCST $T = \{0a, ab, 0c, cd, de\}$ with total cost 6. According to the serial rule, these costs are allocated as $x^S(T) = (1, 3, 0.67, 0.67, 0.67)$. Assume that the link cost k_{ab} is increased to ∞. This changes the MCST, which now becomes

$$T' = \{0a, 0c, cd, de, eb\}$$

with total cost 6.1. According to the serial principle, these costs are allocated as $x^S(T') = (2, 2.6, 0.5, 0.5, 0.5)$, violating cost monotonicity for agent b.

It is also easy to demonstrate that ϕ^B and ϕ^{DK} violate population monotonicity. Yet there is a rule that satisfies the core constraints, strong monotonicity, and population monotonicity: the Folk solution, given by (2.32).

Bergantiños and Vidal-Puga (2007) show that the Folk solution is the only rule satisfying strong monotonicity, population monotonicity, and a property they call "equal share of extra cost." This property states that if all agents have the same connection cost to the root, and this cost is higher than any other connection cost in the network, any increase in the connection cost to the root should be shared equally between all agents (as only one agent should connect to the root, and this agent can be chosen at random).

Consider another version of such a property.

Constant share of extra costs   Let K and K' be two cost matrices where all agents have the same connection cost to the root, which is maximal among all connection costs. Abusing notation, let $K + x$ be the matrix K to which the cost to the root has been increased by the amount x. Then we have

$$\phi(N, K + x) - \phi(N, K) = \phi(N, K' + x) - \phi(N, K').$$

That is, if an amount x is added in both problems, then ϕ should allocate that extra amount in the same way in both networks.

As shown in Bergantiños and Kar (2010) this axiom, together with strong monotonicity and population monotonicity, characterizes the family of obligation rules defined in (2.27).

Theorem 2.5 ϕ satisfies strong cost monotonicity, population monotonicity, and constant share of extra cost if and only if ϕ is an obligation rule, that is, $\phi = \phi^o$.

Although the Folk solution satisfies several desirable properties, the fact that it disregards a lot of cost information (by focusing on the irreducible cost matrix) has lead to criticism. Arguments against so-called reductionist rules are put forward in Bogomolnaia and Moulin (2010). For example, consider a case with ten agents where $k_{0i} = 10$ for all i; $k_{ij} = 1$ for all $i, j \geq 2$; and $k_{1i} = 0$ for all $i \geq 2$. Hence, any MCST will form a star with agent 1 in the middle and some arbitrary agent connected to the root. Note that the presence of agent 1 reduces the total cost for the remaining agents from 18 to 10, so it seems natural that agent 1 should be charged less than the other agents. Yet using the Folk solution, the total cost of 10 will be split equally among the ten agents.

In MCST problems, where the cost of connecting any two agents is less than that of connecting any agent to the root, it can therefore be argued that some sort of strict ranking property should apply. For example, if for all $i, j \in N$, we have that $k_{iz} < k_{jz}$ for all $z \in N \setminus \{i, j\}$ and $k_{0i} \leq k_{0j}$, then $\phi_i(N, K) < \phi_j(N, K)$.

It is easy to see that no reductionist rule (i.e., a rule based on the irreducible matrix) can satisfy such a strict ranking property. In the example mentioned in the previous paragraph, notice that the irreducible matrix is given by $k_{0i}^* = 10$ for all $i \in N$, and $k_{ij}^* = 0$ for all $i, j \in N$. Hence, k_{ij} for all $i, j \geq 2$ can vary from 0 to ∞ without affecting the irreducible matrix. But for any positive value, strict ranking requires that agent 1 pays strictly less than the other agents.

This section ends by providing a characterization of the Shapley value of the game (N, c) defined in (2.28). Recall that for every problem (N, K), there exists a uniquely defined game (N, c). The Shapley value satisfies cost monotonicity and strict ranking (as defined above), but it satisfies neither strong cost monotonicity nor population monotonicity.

It can be argued that no agent should subsidize other agents in case the cheapest way to connect all agents to the root is via individual connections (corresponding to the case where the MCST is a star with the root as its center). Formally, we have the following.

No cross subsidization If the MCST is a star, then $\phi_i(N, K) = k_{0i}$ for all $i \in N$.

This is a mild condition satisfied by all rules mentioned so far.

Moreover, when we consider MCSTs that consists of separate trees (each connected to the root), the same line of argument would suggest that changing the edge cost between any two agents associated with the same tree should not alter the cost shares of agents associated with other trees. To be more precise, we need additional definitions.

An edge ij is called *relevant* if $k_{ij} \leq \max\{k_{0i}, k_{0j}\}$. A path from i to j is called relevant if every edge on the path is relevant. Hence, for a given edge-cost matrix K, agents can be partitioned into groups $N = \{N_1, \ldots, N_p\}$, where members of a given group are connected by some relevant path, while there are no relevant paths between members of different groups.

Group independence   Suppose that some partitioning

$$N = \{N_1, \ldots, N_p\}$$

occurs for cost matrix K. Consider another cost matrix K', where $k'_{nm} = k_{nm}$ for all $\{n, m\} \neq \{i, j\}$, where $i, j \in N_l$. Then $\phi_k(N, K) = \phi_k(N, K')$ for all $k \in N_t$ and for all $t \neq l$.

Matrices K and K' are identical except (perhaps) for the edge cost k_{ij}. Note that when the edge cost k_{ij}, for $i, j \in N_l$, changes, then perhaps the group N_l is further partitioned, but the remaining groups are unchanged. In terms of the allocation problem, the change in edge cost may give rise to changes in the cost of coalitions $S \supseteq \{i, j\}$.

Considering the agents $i, j \in N$ for which the edge cost is changed, it can also be argued that these agents ought to be affected in the same way (i.e., either they both gain or lose the same amount). This property is formally stated as follows.

Equal treatment   Consider two cost matrices K and K', where $k'_{nm} = k_{nm}$ for all $\{n, m\} \neq \{i, j\}$. Then $\phi_i(N, K') - \phi_i(N, K) = \phi_j(N, K') - \phi_j(N, K)$.

As shown by Kar (2002), these three conditions uniquely characterize the Shapley value.

Theorem 2.6 ϕ satisfies no cross subsidization, group independence, and equal treatment if and only if it is the Shapley value ϕ^{Sh}.

Sketch of proof: The Shapley value satisfies all three axioms. Following Kar (2002), we now demonstrate that there is a unique rule satisfying all three axioms, done by induction on the number of relevant edges.

If there are no relevant edges, the MCST is a star and by no cross subsidization, there is a unique solution $\phi_i = k_{0i}$ for all i. Assume that there is a unique solution for all matrices K with at most $(k - 1)$ relevant edges, and consider a matrix K^k with k relevant edges.

By contradiction, assume that there exists a solution $\gamma(N, K^k) \neq \phi(N, K^k)$. Moreover, assume that going from cost matrix K^{k-1} to matrix K^k, only the edge $\{n, m\}$ is made relevant, with $n, m \in N_t$ of the related partition with respect to K^k.

By group independence, $\phi_i(N, K^k) = \gamma_i(N, K^k)$ for all $i \in N_l$, where $l \neq t$. For a relevant edge in N_t we have by the equal treatment property and the induction hypothesis that

$$\phi_m(N, K^k) - \phi_n(N, K^k) = \phi_m(N, K^{k-1}) - \phi_n(N, K^{k-1})$$

$$= \gamma_m(N, K^{k-1}) - \gamma_n(N, K^{k-1}) = \gamma_m(N, K^k) - \gamma_n(N, K^k).$$

Since there are relevant edges between all agents in N_t, we have that

$$\phi_i(N, K^k) - \gamma_i(N, K^k) = \phi_j(N, K^k) - \gamma_j(N, K^k) \quad \text{for any } i, j \in N_t.$$

Since $\sum_{i \in N}(\phi_i(N, K^k) - \gamma_i(N, K^k)) = 0$ and $\phi_i(N, K^k) = \gamma_i(N, K^k)$ for all $i \in N_l$, where $l \neq t$, we get that $\sum_{i \in N_t}(\phi_i(N, K^k) - \gamma_i(N, K^k)) = 0$, and since

$$\phi_i(N, K^k) - \gamma_i(N, K^k) = \phi_j(N, K^k) - \gamma_j(N, K^k) \quad \text{for any } i, j \in N_t,$$

we have in particular that $\phi_i(N, K^k) = \gamma_i(N, K^k)$ for all $i \in N_t$ and consequently that $\phi(N, K^k) = \gamma(N, K^k)$, providing the contradiction. ∎

2.3 Model Variations

There is a rich literature on variations of the MCST model. For example, Farley et al. (2000) consider multiple sources, Suijs (2003) considers costs with both a deterministic and a stochastic element, Bergantiños and Gomez-Rua (2010) consider groups of agents, and Dutta and Mishra (2012) consider asymmetric costs. For the remainder of this chapter, we focus on three variations: costs depending on usage (congestion) in section 2.3.1, decentralized allocation rules (abandoning budget-balance) in section 2.3.2, and introduction of public nodes (Steiner nodes) in section 2.3.3.

2.3.1 Congestion

When edge costs are no longer constant but depend on usage, the edges stop being public goods. This not only complicates the task of finding an efficient network, it also challenges fair allocation.

Quant et al. (2006) and Kleppe et al. (2010) study this variation under the name *congestion network problems*. In practice, there may be several reasons for congestion. For instance, when several agents need a given connection, it typically requires a higher capacity and therefore the cost of using the connection will increase.

Consider first the simple (linear) case, where the cost of using an edge for k agents is simply k times the cost of a single agent. Let $c_{ij}(k)$ denote the cost of using edge ij for k agents (capacity k). Then in the linear case, we have that $c_{ij}(k) = kc_{ij}(1)$, for all edges ij.

For each agent there is a (not necessarily unique) cost-minimizing path to the source. It is intuitively easy to see that if costs are linear in the above sense, then the optimal spanning tree is found as the union of minimum-cost paths of all the agents. Thus, the total minimum cost is found as the sum of edge costs in the optimal tree multiplied by the number of agents using the different edges.

In terms of cost allocation, it is therefore straightforward to suggest that each agent pays the cost of her cost-minimizing path to the source, coinciding with use of the serial principle defined in (2.16). Consider the following three-agent example.

Example 2.10 (Linear congestion costs) Let $N = \{a, b, c\}$, and let the problem be given by the complete graph below.

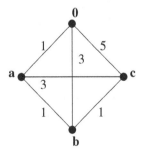

In the figure, edge costs represent the cost of single-agent use $c_{ij}(1)$. Assume that edge costs are linear, such that $c_{ij}(k) = kc_{ij}(1)$ for all edges ij and number of agents $k \in \{1, 2, 3\}$. Clearly, the cost-minimizing path to the source for agent a is the direct connection with cost 1. For agent b, the cost-minimizing path goes via a with a total cost of $1 + 1 = 2$. For agent c, the cost-minimizing path to the source is via a and b with a total cost of $1 + 1 + 1 = 3$. Thus, the optimal graph becomes $\{0a, ab, bc\}$ with a total cost of $1 \times 3 + 1 \times 2 + 1 \times 1 = 6$. Applying the serial principle, costs are shared as $(x_a, x_b, x_c) = (1, 2, 3)$, corresponding to the costs of each agent's cost-minimizing path to the source.

Now consider the problem as a game. First, assume that nodes are *private*, in the sense that agents in $S \subseteq N$ can only connect to the root using nodes in S^0. Thus, we have the game

$$c(a) = 1, c(b) = 3, c(c) = 5,$$
$$c(a, b) = 3, c(a, c) = 5, c(b, c) = 7,$$
$$c(a, b, c) = 6.$$

So the core contains the serial allocation and, in this case, even allows for transfers to agent a (e.g., using the core allocation $(-1, 3, 4)$), since a's cheap connection to the root benefits all coalitions.

Second, assume that nodes are *public*, in the sense that agents in $S \subseteq N$ can freely connect to the root using all the nodes in N^0. Thus, we have the game

$$\bar{c}(a) = 1, \quad \bar{c}(b) = 2, \quad \bar{c}(c) = 3,$$
$$\bar{c}(a, b) = 3, \quad \bar{c}(a, c) = 4, \quad \bar{c}(b, c) = 5,$$
$$\bar{c}(a, b, c) = 6.$$

Notice that this game is *additive* (in the sense that $\bar{c}(S) = \sum_{i \in S} \bar{c}(i)$ for all $S \subseteq N$), so the marginal cost of any agent i, for joining any coalition $S \in N \setminus \{i\}$, is exactly the cost of i's cost-minimizing path to the root. The Shapley value of the (congestion) game with public nodes, $\bar{c}$, therefore coincides with the result of the serial rule (and the unique core element in this case).

Comparing with the game c (with private nodes), we get the Shapley value $(0, 2, 4)$, which is a core element.

Since additive games are (totally) balanced, the core of the congestion game with public nodes, $\bar{c}$, is nonempty (clearly the Shapley value, which equals the serial allocation in this case, satisfies the core conditions). The core of the congestion game with private nodes, c, is nonempty too. Indeed, the serial principle generates a core allocation also in this case: using the serial rule, all agents pay the cost of their cost-minimizing path to the source, say, x_i. For an arbitrary subset of agents $S \subset N$, let $\hat{x}_i$ be the cost of the cheapest path of agent i to the source in the problem $(S^0, c_{|S})$. Clearly, $\sum_{i \in S} \hat{x}_i = c(S)$. Consequently, $\sum_{i \in S} x_i \leq \sum_{i \in S} \hat{x}_i = c(S)$, as desired. Recall that in the standard MCST model without congestion, the cost allocation of the serial rule may violate the core conditions (cf., example 2.6).

More generally, congestion may lead to *concave costs* in the sense that $c_{ij}(k+1) - c_{ij}(k) \leq c_{ij}(k) - c_{ij}(k-1)$ for all ij and $k \in \{1, \ldots, n-1\}$, or *convex costs* in the sense that $c_{ij}(k+1) - c_{ij}(k) \geq c_{ij}(k) - c_{ij}(k-1)$ for all ij and $k \in \{1, \ldots, n-1\}$.

In Quant et al. (2006), it is shown that with convex costs, the core of the (congestion) game with private nodes, c, is nonempty, while there may be instances where the core is empty with concave costs. For the congestion game with public nodes, $\bar{c}$, the core may be empty for both concave and convex congestion costs.[14]

Finding a minimum-cost graph also becomes much more complex than in the linear case. While optimal graphs under concave congestion costs remain trees, convex congestion costs may lead to optimal graphs with cycles.

Example 2.10 (continued: Convex congestion costs)   Now assume that the cost of edges $0a$ and $0b$ are no longer linear in the number of agents, but are convex and given by, $c_{0a}(2) = 3$, $c_{0a}(3) = 6$ and $c_{0b}(2) = 7$, $c_{0b}(3) = 11$. This changes the problem fundamentally. Using the "old" optimal spanning tree, we now get a total cost of $c_{bc}(1) + c_{ab}(2) + c_{0a}(3) = 1 + 2 + 6 = 9$, which is no longer minimal: connecting c to the source via b and a, and a through its direct link is still the cheapest option for a and c, but for b it now becomes cheaper to connect b directly to the source (since sending three agents through $0a$ is now too expensive compared to the linear case), because the total cost is $c_{0a}(2) + c_{0b}(1) + c_{ab}(1) + c_{bc}(1) = 8$. Thus, the optimal graph includes a cycle, as illustrated below.

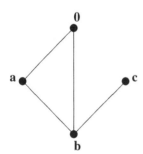

This time the game with private nodes becomes

$$c(a) = 1, \quad c(b) = 3, \quad c(c) = 5,$$
$$c(a, b) = 4, \quad c(a, c) = 6, \quad c(b, c) = 8,$$
$$c(a, b, c) = 8,$$

while the game with public nodes becomes

$$\bar{c}(a) = 1, \quad \bar{c}(b) = 2, \quad \bar{c}(c) = 3,$$

14. Trudeau (2009) presents a subset of concave problems for which the core is nonempty.

$\bar{c}(a,b) = 4, \quad \bar{c}(a,c) = 5, \quad \bar{c}(b,c) = 6,$

$\bar{c}(a,b,c) = 8.$

The core of the game c is nonempty and consists of all allocations $x_a + x_b + x_c = 8$ for which $0 \le x_a \le 1$, $2 \le x_b \le 3$, and $4 \le x_c \le 5$. However, the core of the game $\bar{c}$ is empty, since, for example, $\bar{c}(a,b,c) - \bar{c}(b,c) = 2 \le x_a \le 1 = \bar{c}(a)$.

So while it seems natural to select an allocation in the core of the game c, it becomes a challenge to find a suitable allocation in case of the game $\bar{c}$. A straightforward suggestion here would be to split the cost of each edge equally among the agents who use this edge when connecting to the root, yielding the allocation $(x_a, x_b, x_c) = (1.5, 3.5, 3)$. Note that this allocation violates the core conditions of the game c, since, for example, agent a is forced to pay more than her standalone cost $c(a) = 1$.

When the core is nonempty, it seems reasonable to select an allocation from the core. However, since the core may be empty, both in the cases of concave and convex congestion costs, we are left with the problem of how to share the cost of efficient networks in such cases. As argued in example 2.10, a straightforward example of an allocation rule with some immediate appeal is the equal liability rule: share the cost of each edge in the efficient graph equally among the agents using this edge to connect to the root. The payment of each agent is then found by summing over all edges in the efficient graph. As shown in Moulin and Laigret (2011) and in Bergantiños et al. (2014), this rule possesses a number of desirable properties.

Bergantiños et al. (2014) further suggest a rule dubbed the *painting* rule.[15] To picture the idea, imagine a multistage process where the agents (nodes) "paint" (or pay for) their path to the root at the same speed. At stage 1, each agent is assigned to the first edge leaving the agent's node toward the root; at stage s, each agent is assigned to the first edge that has not been completely paid for, and so forth, until every edge is paid for. However, note that this rule only works for trees and hence cannot be used in cases like example 2.10 (with convex costs).

In Kleppe et al. (2010), it is argued that with public nodes, it becomes necessary to consider two types of associated games: a *direct cost game*, where each coalition S minimizes its cost as if the complement $N \setminus S$ does not make use of any of the edges; and a *marginal cost game*, where each coalition S minimizes its cost as if the complement $N \setminus S$ already is connected to the source

---

15. For the precise but cumbersome definition, see Bergantiños et al. (2014), pp. 782–783.

and therefore makes use of some of the edges. Note that when congestion costs are linear, these two games coincide. The direct cost game seems most relevant when congestion costs are concave, since coalitions should not have an advantage from other agents already connected to the source, while the reverse argument seems to favor the marginal cost game when congestion costs are convex.

Kleppe et al. (2010) show that for concave congestion costs, we can still have games with an empty core when nodes are public, so they focus on convex congestion costs. Kleppe et al. (2010) present an algorithm to determine the optimal network for each coalition of agents and show that the marginal cost game is concave. It is therefore natural to use allocation rules like the Shapley value, but Kleppe et al. (2010) further suggest a core selection that is related to the optimal network itself.

2.3.2   Individual Guarantees and Decentralized (Pricing) Rules

When agents join a network, they know that their benefits and costs will depend on the other agents' characteristics in the network. Yet individual guarantees, here in the form of an upper bound on payments, have always been a crucial issue in the literature on fair allocation (see, e.g., Steinhaus, 1948). The standalone upper bound, $sa(k_{0i}, k[i]) = k_{0i}$, is a classic example of such an individual guarantee, where the sum over all agents covers at least the total cost of the network. As an allocation rule, it is not necessarily budget balanced, but it typically produces a large surplus.

Rules like the standalone rule sa will therefore be dubbed *decentralized pricing rules*, where "decentralization" refers to the fact that agent i's payment only depends on agent i's characteristics, that is, k_{0i} and $k[i] = \{k_{ij}\}_{j \in N \setminus i}$. In practice, decentralized pricing rules have the advantage that an agent can compute his contribution to the network even with the partial information of his own connection costs.

It is easy to see that decentralized rules are in conflict with budget balance. Indeed, consider a problem where $k_{ij} = 0$ for all $i, j \in N^0$, resulting in payments of 0 for everyone. Now change the connection cost to the root for one agent from 0 to 1. When cost shares are decentralized, none of the other agents change their payment, and by budget balance the last agent's payment is not changed either. Sequentially continuing this operation we end up in a case where all agents have $k_{0i} = 1$, and yet $n - 1$ agents are paying 0 and a single agent is paying 1 (by budget balance) in a symmetric problem: a contradiction.

Thus, decentralized rules produce a budget surplus, and typically even a substantial one, as in the case of the standalone rule. So the question is whether decentralized pricing rules exist for which the surplus is sufficiently small while satisfying other relevant and desirable properties.

Hougaard et al. (2010) analyze a particular decentralized pricing rule that satisfies a number of desirable properties as well as incur a relatively low surplus, dubbed the *canonical pricing rule*. It is defined as follows.

Let Π_N be the set of all orders of N. Given $\pi \in \Pi_N$, let $\mathcal{P}(i, \pi)$ denote the union of the root and the set of agents prior to agent i in the order π: $\mathcal{P}(i, \pi) = \{0\} \cup \{j \in N | \pi(j) < \pi(i)\}$. For each agent $i \in N$, the canonical pricing rule is given by

$$can(k_{0i}, k[i]) = \frac{1}{n!} \sum_{\pi \in \Pi_N} \min_{j \in \mathcal{P}(i,\pi)} \{k_{ij}\}. \tag{2.34}$$

Equation (2.34) provides a simple interpretation of the rule. Pick a given ordering of the agents. Let the first agent pay the connection to the root; let the second agent pay her cheapest connection either to the root or to agent 1;... ; let the jth agent pay her cheapest connection cost to the root or to any of the predecessors of j; and so forth. Do this for every ordering, and take the average over orderings for each agent.

Note that for each ordering, this idea is similar to the Prim algorithm but with the important difference that the ordering is fixed a priori.

Obviously the expression in (2.34) is not very fit for computation of payments. For this use we can rewrite the definition as follows,

$$can(k_{0i}, k[i]) = \frac{1}{n}k_{0i} + \sum_{t=1}^{n-1} \frac{1}{t(t+1)} \min\{k_{0i}, k_i^t\}, \tag{2.35}$$

where we arrange the $n - 1$ numbers $k[i]$ increasingly as k_i^t, $1 \le t \le n - 1$ so that $k_i^1 \le \cdots \le k_i^{n-1}$.

Example 2.6 (continued: Illustrating the canonical pricing rule and its relation to the Folk solution) Recall example 2.6, where the problem is given by the following complete graph.

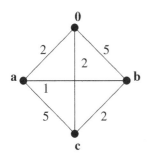

The standalone pricing rule results in the allocation $sa = (2, 5, 2)$ for agents a, b, and c, respectively. This allocation runs a surplus of 4 (compared to the minimal cost of 5).

Consider the canonical pricing rule: $k[a] = (1, 5)$, $k[b] = (1, 2)$, and $k[c] = (2, 5)$, so using the expression (2.35) we have

$$can(k_{0a}, k[a]) = 1.5,$$

$$can(k_{0b}, k[b]) = 2.5,$$

$$can(k_{0c}, k[c]) = 2.$$

Thus, the resulting surplus is 1 (compared to the minimum cost of 5). So the surplus is considerably reduced compared to the standalone rule.

Looking at the irreducible costs, the problem is reduced to the following.

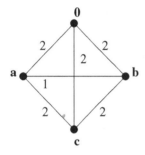

In this case, we have $k[a] = (1, 2)$, $k[b] = (1, 2)$, and $k[c] = (2, 2)$, with

$$can(k_{0a}, k[a]) = 1.5,$$

$$can(k_{0b}, k[b]) = 1.5,$$

$$can(k_{0c}, k[c]) = 2.$$

Note that the surplus is 0, and the result of the canonical rule coincides with the Folk solution (or equal remaining obligations rule) in this case. In fact, this turns out to be true in general:

$$can(k_{0i}^*, k^*[i]) = \phi_i^{Folk}(N, K),$$

for all N, K, and i. This also implies that the Folk solution is always bounded from above by the canonical rule.

It can be shown that the canonical pricing rule satisfies both cost monotonicity and population monotonicity. Moreover, the rule is the smallest among those improving on the standalone upper bound satisfying the structural property of superadditivity (i.e., for any pair $(k_{0i}^1, k^1[i])$ and $(k_{0i}^2, k^2[i])$, we have $\psi(k_{0i}^1 + k_{0i}^2, k^1[i] + k^2[i]) \geq \psi(k_{0i}^1, k^1[i]) + \psi(k_{0i}^2, k^2[i]))$.[16] Further axiomatic characterizations can be found in Hougaard et al. (2010).[17]

2.3.3 Minimum-Cost Steiner Trees

Another important variation of the MCST model allows for the presence of public nodes: so-called *Steiner nodes*. These nodes do not represent agents but can be accessed by any agent when connecting to the root.

In its original form, the Steiner problem is geometrical: determine a set of edges of minimum length connecting a given set of nodes in the Euclidian plane. For example, consider the figure (a) below, where two agents A and B are connected to a root 0 in a chain and costs of edges are proportional to the length. By adding a Steiner node, S, as in figure (b), the total length of the graph can be reduced.

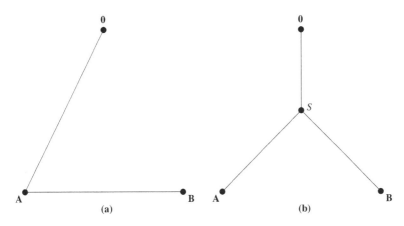

The need to solve various variations of Steiner problems has cropped up in many practical applications in different types of industries, ranging from design of very-large-scale integration, creating integrated circuits, to network routing (see, e.g., Cheng et al. 2004). With particular relevance for cost allocation, we may think of designing transportation networks or backbone networks of cables or pipes, using the introduction of Steiner nodes.

In general, edge costs do not need to be constrained by the Euclidian metric, so in relation to cost allocation (of the resulting optimal Steiner tree), we may formulate the *minimum-cost Steiner tree problem* as follows. Given

---

16. Superadditivity can be seen as the planner having preference for flexibility in the sense that it is cheaper to implement two separate solutions than the joint solution.

17. See exercise 2.10.

a cost-weighted (undirected) graph $G^0 = (N \cup \{0\}, E)$ and a set of nodes (agents) $A \subseteq N$, find the cost-minimizing tree in G^0 with a node set containing $A \cup \{0\}$. Allocate the total minimal cost among the set A.

Clearly, if $A = N$, we are back to the MCST model. Yet, contrary to the MCST model, the minimum-cost Steiner tree model may have an empty core, as first noted in Megiddo (1978b). The following elegant three-agent example is due to Tamir (1991).

Example 2.11 (Minimum-cost Steiner tree problems may have empty cores)
Consider the network below with root 0 and agents $\{a, b, c\}$, where all edges have cost 1.

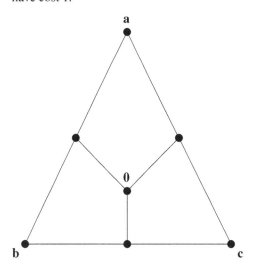

The minimum-cost Steiner tree connecting all three agents to the root needs five edges (and thereby has a total cost of 5). Any two-agent coalition needs three edges (and thereby has a total cost of 3), while single-agent coalitions need two edges (and thereby have a total cost of 2). Consequently, the core conditions imply that the cost share of any agent must be 2 (since $c(ijh) - c(jh) = 2 \le x_i \le c(i) = 2$). However, cost shares must sum to 5, so the core is empty.

Introducing Steiner nodes in the MCST problem has a substantial impact on cost allocation issues, not only because it introduces the possibility of an empty core, but also because such "public" nodes further complicate the cost externalities involved. The following simple two agent, one Steiner node example illustrates this point.

Example 2.12 (Steiner nodes and their effect on fair allocation) First, consider a situation described by the graph below. Ann and Bob demand connection to the root 0, and edge costs are given as in the graph. S is a Steiner node.

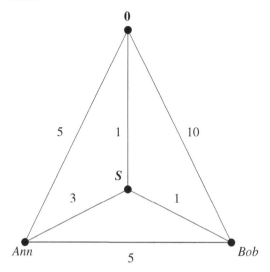

Disregarding the Steiner node, S, the situation seems fairly simple: Bob has twice as high standalone costs as Ann has, and the cost-minimizing tree is $T = \{01, 12\}$ with total cost 10. If each agent should pay for their connecting edge in the efficient graph, they would split the cost equally (as if using the Bird rule). Yet Bob saves 5 compared to his standalone cost, and it seems reasonable that his cost saving of 5 should be split equally between them, resulting in the allocation $(2.5, 7.5)$: this is indeed the Shapley value of the associated MCST game as well as the result of applying the serial principle on the MCST $T = \{01, 12\}$.

Adding the Steiner node clearly changes the efficient network. It now connects both agents to the root via the Steiner node with a total cost of 5. Using the proportions from before (based on their standalone cost and disregarding the Steiner node), we therefore have the allocation $(\frac{5}{3}, \frac{10}{3})$. However, with access to the Steiner node, the standalone cost of the two agents has changed to 4, and 2, respectively, indicating that this line of reasoning should now lead to the allocation $(\frac{10}{3}, \frac{5}{3})$, which may seem more compelling.

Yet another straightforward approach to cost sharing in this case would be the following. To reach the Steiner node, agents have individual costs of 3 and 1, respectively. As they both use the connection from the Steiner node to the root, this cost of 1 should be split equally between them, resulting in the final allocation $(3.5, 1.5)$. The challenge is therefore how to balance these different considerations of fairness in a given allocation rule.

In cases like this one, where the addition of a Steiner node makes it possible to establish a cheaper network than before, it is tempting to disregard the original edge costs (k_{01}, k_{02}, k_{12}) when splitting the total cost of the new graph, since they are not relevant for the resulting network anyway. Likewise, if adding the Steiner node does not influence the efficient network of the original problem, the cost of edges related to the Steiner node should be disregarded when allocating the total cost. In fact, in terms of implementation (to which we return in chapter 6) only the costs of the resulting network will be realized, and it is questionable whether the estimated cost of links that are not realized should influence the allocation of the actual costs. If this is the case, agents have an incentive to manipulate the estimated cost of nonrealized links to the extent that this is possible.

Second, when the addition of a Steiner node makes it possible to form a network for the same cost as the original efficient network (e.g., as in the case illustrated below), it can be argued that the allocation of the total cost should not depend on which of the two options is realized.

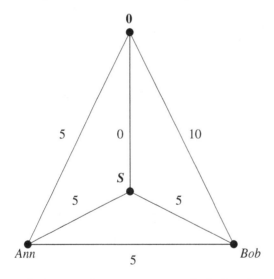

Such a requirement seems particularly relevant in terms of implementation, while it may clash with the perceptions of fairness mentioned above. Indeed, looking at the network involving the Steiner node, splitting the total cost equally seems the most reasonable solution, that is, $(5, 5)$. This can be compared to the Shapley value of the original problem (disregarding the Steiner node), which is $(2.5, 7.5)$. In the context of minimum cost connection networks, we will return to such considerations in chapter 4 and 6.

2.4 Exercises

2.1. (Hougaard et al., 2017) Consider the linear hierarchy revenue-sharing problem. Prove that the no-transfer rule (2.1) can be characterized by highest-rank revenue independence and a property called revenue order preservation: For each (N, r), and each pair $i, j \in N$ such that $r_i \geq r_j$, $\phi_i(N, r) \geq \phi_j(N, r)$.

Also, prove that the full-transfer rule (2.2) can be characterized by highest-rank splitting neutrality and a property called hierarchical order preservation: For each (N, r), and each pair $i, j \in N$ such that $i \geq j$, $\phi_i(N, r) \geq \phi_j(N, r)$.

Finally, consider the following condition (which is called canonical fairness in Hougaard et al. 2017): For each $x \in \mathbb{R}_+$, $\phi(\{1, 2\}, (x, 0)) = (\frac{x}{2}, \frac{x}{2})$. Comment on this condition, and find the set of axioms from the characterization in theorem 2.1, which together with canonical fairness characterizes the intermediate rule $\phi^{0.5}$.

2.2. Consider again the linear hierarchy revenue-sharing problem (N, r), and assume that the boss n is free to decide the size of the hierarchy, while groups of subordinates have no value on their own (e.g., the boss is a patent owner). That is, we may represent the associated cooperative game by the value function

$$v(S) = \begin{cases} \sum_{j \in S} r_j & \text{if } n \in S \\ 0 & \text{otherwise} \end{cases}$$

for all $S \subseteq N$. Compute the Shapley value of the game (N, v), and compare with the Shapley value of the additive revenue game restricted by the linear hierarchy permission structure given by (2.6). Moreover, compare these results to that of using a fixed-fraction transfer rule (2.3).

2.3. (Ambec and Sprumont, 2002) Recall the river-sharing model in section 2.1.1. Prove that $z^{DI} \in \text{core}(N, v)$. (Hint: use that v is superadditive.) Moreover, it can be shown that the game (N, w) is concave. Prove that $z^{DI} \in \text{core}(N, w)$. Suggest some allocation rules that are core selections in the game (N, v) but not in game (N, w), and vice versa.

2.4. (Hougaard et al., 2017) Consider the family of sequential λ-contributions rules defined in (5.11) and (2.19) for cost sharing in standard fixed trees. Illustrate how payments are determined when $\lambda = 0$.

Define the corresponding family of rules for a branch hierarchy revenue-sharing problem. Find an axiomatic characterization of this family. (Hint: Use suitable versions of the axioms for the linear case in section 2.1.1.)

2.5. Consider the MCST problem with root 0 and three agents $\{a, b, c\}$ illustrated below, with ε close to zero.

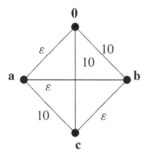

Formulate the MCST problem as a game (N, c) (see definition (2.28)). Find the core, the Shapley value, and the Lorenz set (i.e., the set of Lorenz-undominated allocations in the core; see Hougaard et al., 2001). Discuss the result. What happens if the edge costs of $0b$, $0c$, and ac are increased from 10 to 100? What happens if the game is defined by (2.29) instead?

2.6. (Arin et al., 2008) Following up on exercise 2.5, a core allocation is called *egalitarian* if it is in the Lorenz set (i.e., if no other allocation in the core Lorenz-dominates it). Consider the class of balanced games, and define some allocation rules that are egalitarian (i.e, result in egalitarian allocations). For example, prove that a rule selecting the core allocation that minimizes the Euclidean distance to the equal-split allocation is egalitarian.

An allocation rule ϕ satisfies *independence of irrelevant core alternatives* if $\phi(N, v) = \phi(N, w)$ for any two games (N, v) and (N, w) with $\phi(N, v) \in core(N, w) \subset core(N, v)$. Show that egalitarian rules satisfy independence of irrelevant core alternatives. Try to prove that every continuous core-allocation rule that satisfies independence of irrelevant core alternatives is egalitarian.

2.7. (Trudeau, 2012) Consider the MCST problem. Let $c^*(N \setminus i)$ be the irreducible cost matrix associated with $(N \setminus i, K_{|N \setminus i})$. Define the *cycle-complete cost matrix* $\check{c}$ such that for $i, j \in N$, $\check{c}_{ij} = \max_{k \in N \setminus \{i, j\}} c_{ij}^*(N \setminus k)$, and for $i \in N$, $\check{c}_{0i} = \max_{k \in N \setminus i} c_{0i}^*(N \setminus k)$. Compute $\check{c}$ for the case in example 2.8. Show that the resulting game $(N, \check{c})$ is concave.

Try to prove that the game $(N, \check{c})$ is concave in general. Moreover, show that $core(N, c^*) \subseteq core(N, \check{c})$.

2.8. Continue example 2.6 for the MCST game (N, c). Compute the Shapley value of this game, and determine whether it is a core allocation. If not, which coalition would block the cooperation? Suppose the planner is able to enforce that any coalition S must share its standalone value using the Shapley value of the projected game for S. Will this potential blocking threat be real? Comment on this issue in general.

2.9. Consider a version of the MCST model where agents $i \in N$ obtain individual benefits $w_i \geq 0$ from being connected to the root (source), and define the welfare game (N, V), where

$$V(S) = \max_{T \subseteq S} \left\{ \sum_{i \in T} w_i - c(T) \right\} \quad \text{for all } S \subseteq N,$$

with $c(T)$ being the minimum cost of connecting coalition T to the root. Note that the network that maximizes the social welfare, $V(N)$, may exclude some agents (simply because the net welfare gain may be lower than the added cost for some agents). Show that the welfare game is superadditive but may have empty core (i.e., it is not balanced). Discuss the implications.

2.10. (Hougaard et al., 2010) Consider the canonical pricing rule (2.35). For a generic pricing rule f, comment on the interpretation, and desirability, of the following two axioms:

Independence of irrelevant links   For any two agents $i, j \in N$ and profile of costs $(k_{0i}, k[i])$,

$$\{k_{ij} \geq k_{0i}\} \implies f(N, k_{0i}, k[i]) = f(N \setminus \{j\}, k_{0i}, k[i, N \setminus \{j\}]).$$

Equal share of extra costs   For any profile of costs $(k_{0i}, k[i])$, agent i, and number $\delta > 0$,

$$\{k_{ij} \leq k_{0i} \text{ for all } j\} \implies f(k_{0i} + \delta, k[i]) = f(k_{0i}, k[i]) + \frac{\delta}{n} \quad \text{for all } i.$$

Show that among rules that are bounded from above by the standalone cost, the canonical pricing rule is uniquely characterized by these two axioms.

2.11. (Trudeau, 2014) Consider a variation of the MCST-problem somewhat close to the minimum-cost Steiner tree problem dubbed a *MCST problem with indifferent agents*. Suppose that the set of agents N can be partitioned into two subsets: those who wants connection to the root (N^d), and those who do not (N^t). Now agents in N^t allow the other agents to pass through their node when connecting to the root. Let (N, c) be the conventional MCST problem (where

$N^t = \emptyset$). Then in the new game with indifferent agents, the cost function is defined as $\hat{c}(S) = \min_{T \subseteq N^t \cap S} c(S \cap N^d \cup T)$ for all $S \subseteq N$. Interpret this game. Show that the core may be empty.

In the special case where $|N^d| = 1$, the problem coincides with the classical *shortest path problem*. Can the core still be empty here?

2.12. Consider the following (fixed) rooted Steiner tree with agents $N = \{a, b, c\}$ and Steiner nodes $S = \{s_1, s_2, s_3\}$.

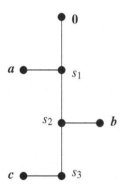

Every edge has cost c_{ij} for $i, j \in N \cup S$. Describe the network structure, and discuss how to allocate the total cost between the agents in N.

References

Aadland, D., and V. Kolpin (1998), "Shared Irrigation Costs: An Empirical and Axiomatic Analysis," *Mathematical Social Sciences* 35: 203–218.

Ambec, S., and L. Ehlers (2008), "Sharing a River among Satiable Agents," *Games and Economic Behavior* 64: 35–50.

Ambec, S., and Y. Sprumont (2002), "Sharing a River," *Journal of Economic Theory* 107: 453–462.

Ambec, S., A. Dinar, and D. McKinney (2013), "Water Sharing Agreements Sustainable to Reduced Flows," *Journal of Environmental Economics and Management* 66: 639–655.

Arin, J., J. Kuipers, and D. Vermeulen (2008), "An Axiomatic Approach to Egalitarianism in TU-Games," *International Journal of Game Theory* 37: 565–580.

Bergantiños, G., and M. Gomez-Rua (2010), "Minimum Cost Spanning Tree Problems with Groups," *Economic Theory* 43: 227–262.

Bergantiños, G., M. Gomez-Rua, N. Llorca, M. Pulido, and J. Sanchez-Soriano (2014), "A New Rule for Source Connection Problems," *European Journal of Operational Research* 234: 780–788.

Bergantiños, G., and A. Kar (2010), "On Obligation Rules for Minimum Cost Spanning Tree Problems," *Games and Economic Behavior* 69: 224–237.

Bergantiños, G., L. Lorenzo, and S. Lorenzo-Freire (2010), "The Family of Cost Monotonic and Cost Additive Rules in Minimum Cost Spanning Tree Problems," *Social Choice and Welfare* 34: 695–710.

Bergantiños, G., and J. J. Vidal-Puga (2007), "A Fair Rule in Minimum Cost Spanning Tree Problems," *Journal of Economic Theory* 137: 326–352.

Bird, C. G. (1976), "On Cost Allocation for a Spanning Tree: A Game Theoretic Approach," *Networks* 6: 335–350.

Bjørndal, E., M. Koster, and S. Tijs (2004), "Weighted Allocation Rules for Standard Fixed Tree Games," *Mathematical Methods of Operations Research* 59: 249–270.

Bogomolnaia, A., and H. Moulin (2010), "Sharing a Minimal Cost Spanning Tree: Beyond the Folk Solution," *Games and Economic Behavior* 69: 238–248.

Branzei, R., S. Moretti, H. Norde, and S. Tijs (2004), "The P-Value for Cost Sharing in Minimum Cost Spanning Tree Situations," *Theory and Decision* 56: 47–61.

Cheng, X., Y. Li, D.-Z. Du, and H. Q. Ngo (2004), "Steiner Trees in Industry," in D.-Z. Du and P.M. Pardalos (eds.,) *Handbook of Combinatorial Optimization* vol. 5. Dordrecht: Kluwer.

Claus, A., and D. J. Kleitman (1973), "Cost Allocation for a Spanning Tree," *Networks* 3: 289–304.

Dubey, P. (1982), "The Shapley Value as Aircraft Landing Fees— Revisited," *Management Science* 28: 869–874.

Dutta, B., and A. Kar (2004), "Cost Monotonicity, Consistency and Minimum Cost Spanning Tree Games," *Games and Economic Behavior* 48: 223–248.

Dutta, B., and D. Mishra (2012), "Minimum Cost Arborescences," *Games and Economic Behavior* 74: 120–143.

Dutta, B., and D. Ray (1989), "A Concept of Egalitarianism under Participation Constraints," *Econometrica* 57: 615–635.

Farley, A. M., P. Frogopoulou, D. Krumme, A. Proskurowski, and D. Richards (2000), "Multi-source Spanning Tree Problems," *Journal of Interconnection Networks* 1: 61–71.

Feltkamp, V., S. Tijs, and S. Muto (1994), "On the Reducible Core and the Equal Remaining Obligation Rule of Minimum Cost Extension Problems," mimeo, Tilburg University, Tilburg, Holland.

Gilles, R. P., and G. Owen (1999), "Cooperative Games and Disjunctive Permission Structures," working paper, Virginia Tech Blacksburgh, VA.

Gilles, R. P., G. Owen, and R. van den Brink (1992), "Games with Permission Structures: The Conjunctive Approach," *International Journal of Game Theory* 20: 277–293.

Granot, D., and G. Huberman (1981), "Minimum Cost Spanning Tree Games," *Mathematical Programming* 21: 1–18.

Granot, D., J. Kuipers, and S. Chopra (2002), "Cost Allocation for a Tree Network with Heterogeneous Customers," *Mathematics of Operations Research* 27: 647–661.

Granot, D., M. Maschler, G. Owen, and W. R. Zhu (1996), "The Kernel/Nucleolus of a Standard Tree Game," *International Journal of Game Theory* 25: 219–244.

Herzog, S., S. Shenker, and D. Estrin (1997), "Sharing the 'Cost' of Multicast Trees: An Axiomatic Analysis," *IEEE/ACM Transactions on Networking* 5: 847–860.

Hougaard, J. L. (2009), *An Introduction to Allocation Rules.* New York: Springer.

Hougaard, J. L., J. D. Moreno-Ternero, M. Tvede, and L. P. Østerdal (2017), "Sharing the Proceeds from a Hierarchical Venture," *Games and Economic Behavior* 102: 98–110.

Hougaard, J. L., H. Moulin, and L. P. Østerdal (2010), "Decentralized Pricing in Minimum Cost Spanning Trees," *Economic Theory* 44: 293–306.

Hougaard, J. L., B. Peleg, and L. Thorlund-Petersen (2001), "On the Set of Lorenz-Maximal Imputations in the Core of a Balanced Game," *International Journal of Game Theory* 30: 147–165.

Hougaard, J. L., and M. Tvede (2012), "Truth-Telling and Nash Equilibria in Minimum Cost Spanning Tree Models," *European Journal of Operational Research* 222: 566–570.

Kar, A. (2002), "Axiomatization of the Shapley Value on Minimum Costs Spanning Tree Games," *Games and Economic Behavior* 38: 265–277.

Kleppe, J., M. Quant, and H. Reijnierse (2010), "Public Congestion Network Situations and Related Games," *Networks* 55: 368–378.

Koster, M., E. Molina, Y. Sprumont, and S. Tijs (2001), "Sharing the Cost of a Network: Core and Core Allocations," *International Journal of Game Theory* 30: 567–599.

Littlechild, S. C. (1974), "A Simple Expression of the Nucleolus in a Special Case," *International Journal of Game Theory* 3: 21–29.

Littlechild, S. C., and G. Owen (1973), "A Simple Expression for the Shapley Value in a Special Case," *Management Science* 20: 370–372.

Littlechild, S. C., and G. F., Thompson (1977), "Aircraft Landing Fees: A Game Theoretic Approach," *Bell Journal of Economics* 8: 186–204.

Lorenzo, L., and S. Lorenzo-Freire (2009), "A Charatcrization of Kruskal Sharing Rules for Minimum Cost Spanning Tree Problems," *International Journal of Game Theory* 38: 107–126.

Megiddo, M. (1978a), "Computational Complexity of the Game Theory Approach to Cost Allocation for a Tree, *Mathematics of Operations Research* 3: 189–196.

Megiddo, N. (1978b), "Cost Allocation for Steiner Trees," *Networks* 8: 1–6.

Miquel, S., B. van Velzen, H. Hamers, and H. Norde (2006), "Fixed Tree Games with Multilocated Players," *Networks* 47: 93–101.

Monderer, D., D. Samet, and L. Shapley (1992), "Weighted Values and the Core," *International Journal of Game Theory* 21: 27–39.

Moulin, H., and F. Laigret (2011), "Equal-Need Sharing of a Network under Connectivity Constraints," *Games and Economic Behavior* 72: 314–320.

Moulin, H., and S. Shenker (1992), "Serial Cost Sharing," *Econometrica* 60: 1009–1037.

Pickard, G., W. Pan, I. Rahwan, M. Cebrian, R. Crane, A. Madan, and A. Pentland (2011), "Time Critical Social Mobilization," *Science* 334: 509–512.

Potters, J., and P. Sudhölter (1999), "Airport Problems and Consistent Allocation Rules," *Mathematical Social Sciences* 38: 83–102.

Quant, M., P. Borm, and H. Reijnierse (2006), "Congestion Network Problems and Related Games," *European Journal of Operational Research* 172: 919–930.

Steinhaus, H. (1948), "The Problem of Fair Division," *Econometrica* 16: 101–104.

Suijs, J. (2003), "Cost Allocation in Spanning Network Enterprises with Stochastic Connection Costs," *Games and Economic Behavior* 42: 156–171.

Tamir, A. (1991), "On the Core of Network Synthesis Games," *Mathematical Programming* 50: 123–135.

Thomson, W. (2013), "Cost Allocation and Airport Problems," Working paper, University of Rochester, Rochester, NY.

Thomson, W. (2015), "Axiomatic and Game-Theoretic Analysis of Bankruptcy and Taxation Problems: An Update," *Mathematical Social Sciences* 74: 41–59.

Tijs, S., R. Branzei, S. Moretti, and H. Norde (2006), "Obligation Rules for Minimum Cost Spanning Tree Situations and Their Monotonicity Properties," *European Journal of Operational Research* 175: 121–134.

Trudeau, C. (2009), "Network Flow Problems and Permutationally Concave Games," *Mathematical Social Sciences* 58: 121–131.

Trudeau, C. (2012), "A New Stable and More Responsive Cost Sharing Solution for Minimum Cost Spanning Tree Problems," *Games and Economic Behavior* 75: 402–412.

Trudeau, C. (2014), "Minimum Cost Spanning Tree Problems with Indifferent Agents," *Games and Economic Behavior* 84: 137–151.

van den Brink, R. (1997), "An Axiomatization of the Disjunctive Permission Value for Games with a Permission Structure," *International Journal of Game Theory* 26: 27–43.

van den Brink, R., and R. P. Gilles (1996), "Axiomatizations of the Conjunctive Permission Value for Games with Permission Structures," *Games and Economic Behavior* 12: 113–126.

van den Brink, R., G. van der Laan, and N. Moes (2012), "Fair Agreements for Sharing International Rivers with Multiple Springs and Externalities," *Journal of Environmental Economics and Management* 63: 388–403.

3 Cycles

Trees often represent efficient network structures, but there may be several reasons for cycles to appear in efficient networks as well. Of course, cycles can arise directly from network activity. For instance, when edges (or flows) in the network are *directed*, as in the digraph below, nodes may represent agents supplying information to each other. For example, agent a delivers information to c, agent c delivers information to b, and agent b delivers information to a, resulting in a directed cycle.

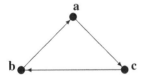

However, cycles can also arise in undirected efficient networks due to the presence of externalities. Recall the congestion network in example 2.10 (section 2.2.4), where a cycle in the efficient network was caused by convex edge costs (i.e., the cost of serving n users via each edge exceeds that of n times the cost of serving a single user).

This phenomenon can arise even for concave edge costs if the congestion network model is generalized. Instead of all wanting to connect to the same node (the root), agents now have individual connection demands in the form of pairs of nodes (a_i, b_i) that they want connected directly or indirectly. Consider the graph below. Suppose that Ann wants to connect nodes a and b, Bob wants to connect nodes b and c, and Carl wants to connect nodes a and c. If edge costs are constant, the cost-minimizing graph is obviously a tree obtained by deleting any one of the three edges. However, if edge costs depend on the capacity (i.e., the number of agents served by the edge) the story is quite different: suppose that edge costs for capacity 1 (that is, for serving one agent) are given by $c_{ij}(1) = 1$, for all $i, j \in \{a, b, c\}$ with $i \neq j$; and edge costs for capacity 2 (that is, for serving two agents) are given by $c_{ij}(2) = 1.75$, for all $i, j \in \{a, b, c\}$

with $i \neq j$. So costs are concave, since serving two agents is less than twice as expensive as serving one agent.

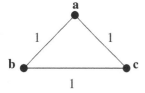

Clearly the cost-minimizing graph is now the entire cycle, because deleting any of the three edges leads to a tree with total cost 3.5, versus the total cost of the cycle being 3.

In this chapter, we focus on classical routing and delivery problems (i.e., fixed route, traveling salesman, and Chinese postman problems), where the cycle is the direct result of the underlying activity itself. In the fixed-route problem the cycle is given by a specific scheduled order of the nodes (beginning and ending with the root), and the scheduled cycle is not necessarily cost minimizing despite full knowledge of all edge costs in the complete graph. In the latter case we need to solve an optimization problem based on full knowledge of edge costs in the complete graph. In the traveling salesman problem, where agents are represented by nodes, we need to find a cost-minimizing (Hamiltonian) cycle for every coalition of agents. In the Chinese postman problem, where agents are represented by edges, we need to find a cost-minimizing (Euler) cycle for every coalition of agents.

In rooted cycles there is no longer a unique path connecting a given node to the root, so agents' network usage is much harder to determine than in the case of tree-structured networks. In some cases it will be relevant to model the allocation problem as a cooperative game and use game-theoretic solution concepts as allocation rules. For instance, this approach is warranted when the cycle is the result of an optimizing procedure, as in the routing problems mentioned above. However, a nonempty core is no longer guaranteed in such games, so certain simple allocation rules may be relevant as well, and they may perhaps even do better in some contexts.

3.1 Fixed Route

Motivating their analysis of fixed-route cost allocation, Fishburn and Pollak (1983, p. 366) start out with the following story:

Not long ago one of us (H.P.) flew from Newark, New Jersey, to Columbus, Ohio, to Raleigh/Durham to Tampa to Washington D.C., then back to Newark. He went to

Columbus on Bell Laboratories business, to Raleigh/Durham for the State of North Carolina, to Tampa for personal reasons and to Washington D.C. on National Science Foundation business. After returning home, he had to decide how much of his air fare to charge each sponsor of his trip, namely, Bell Labs, the State of North Carolina, himself, and the NSF. He wished to do this in as fair a manner as possible. At the minimum, he felt that the sponsors should be charged exactly the total air fare, and that no sponsor should pay more than the direct round-trip cost between Newark and the sponsor's city.

An important feature here is that the chosen route (directed cycle) may not be cost minimizing but is selected for all kinds of other reasons (e.g., the order of scheduled business meetings). Clearly, this differs fundamentally from the usual type of network design problems, where the aim is to find the efficient network (here a cost-minimizing Hamiltonian cycle). The conventional game-theoretic approach is still valid, but the core may be empty, and standard allocation rules like the Shapley value may not even satisfy individual rationality (in the sense that sponsors may be charged more than the cost of a direct round trip from home to the sponsor's city). Thus, there are good reasons to search for much simpler allocation methods that at least ensure individual rationality, for example, a round-trip version of the proportional rule.

3.1.1 The Model

Assume that an agent (e.g., a person or a vehicle) travels from the home city (root) 0 to n sponsor cities in the order $\pi = 1, 2, \ldots, n$, and then returns home. Let $N = \{1, \ldots, n\}$ denote the set of sponsor cities, and let Π be the set of all $n!$ possible orders of the n sponsor cities. Let $c_{ij} > 0$ denote the (constant) cost of a direct trip from city i to j ($i \neq j$) with $i, j \in N^0 = N \cup \{0\}$. Moreover, assume that costs are symmetric ($c_{ij} = c_{ji}$), and that $c_{ii} = 0$ for all i.

Assume further that costs satisfy the triangular inequalities with respect to the home city, such that the cost of a direct trip between cities i and j never exceeds that of going via the home city 0:

$$c_{0i} + c_{0j} \geq c_{ij}, \tag{3.1}$$

for all $i, j \in N$.

Let C^π denote the total cost of the (ordered) tour $\pi \in \Pi$. Note that, because of (3.1), we have that $C^\pi \leq \sum_{i \in N} 2c_{0i}$, for any order π. (Indeed, for any adjacent nodes on the tour, $c_{i,i+1} \leq c_{0i} + c_{0,i+1}$, and therefore adding over all cities and adding the trip from 0 to the first city, as well as the last trip from city n to 0, we get $c_{01} + c_{12} + \cdots + c_{0n} \leq c_{01} + (c_{01} + c_{02}) + (c_{02} + c_{03}) + \cdots + (c_{0,n-1} + c_{0n}) + c_{0n} = \sum_{i \in N} 2c_{0i}$).

The total cost C^π of route π has to be allocated among the n sponsor cities using a budget-balanced cost-allocation rule ϕ: $\sum_{i \in N} \phi_i = C^\pi$.

3.1.2 Fixed-Route Game

Even though the tour represents a fixed network, it is straightforward to define an associated game, given the order of the sponsor cities π. For all singleton city coalitions $i \in N$, $c^{\pi}(i) = 2c_{0i}$. For any coalition $S \subset N$ of cities, (possibly relabeled $\{(1), \ldots, (s)\}$ and ordered according to the original order π), $c^{\pi}(S) = c_{0(1)} + c_{(1)(2)} + \cdots + c_{(s-1)(s)} + c_{0(s)}$, with $c^{\pi}(N) = C^{\pi}$. The pair,

$$(N, c^{\pi}), \tag{3.2}$$

is dubbed a *fixed-route cost-allocation game*.

Derks and Kuipers (1997) show that the core of the fixed-route cost-allocation game (3.2) is nonempty if and only if $c^{\pi}(N) \leq c^{\pi}(S) + c^{\pi}(N \setminus S)$ for all $S \subseteq N$: that is, it never pays off to split the tour into (two) sub-tours.

Yet, as demonstrated by example 3.1 below, it is easy to show that the game defined by (3.2) may have an empty core, and that standard game-theoretic solutions like the Shapley value may violate even basic requirements, such as individual rationality, which in this case states that $x_i \leq 2c_{0i}$ for all $i \in N$.

Example 3.1 (The game (N, c^{π}) can have an empty core and the Shapley value may violate individual rationality) Consider connection costs between three cities $i \in \{1, 2, 3\}$, and a home city 0, given by the complete graph below.

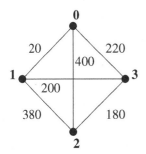

Clearly, the chosen route (order of the cities visited) is important for the total cost of the tour. If the sponsor cities are visited in the orders $\pi \in \{(213), (312)\}$ we get a total cost of $C^{\pi} = 1200$, while for any of the remaining orders $\pi \in \{(123), (132), (231), (321)\}$, the total cost is only $C^{\pi} = 800$. Thus, for a given problem, the total cost is not necessarily minimized, and visiting the cities in a different order may be cheaper.

For the order $\pi = \{2, 1, 3\}$, with total cost $C^{\pi} = 1200$, we get the following game:

$$c^{\pi}(1) = 40, \; c^{\pi}(2) = 800, \; c^{\pi}(3) = 440;$$

$$c^{\pi}(12) = c^{\pi}(23) = 800, \; c^{\pi}(13) = 440;$$

$$c^{\pi}(123) = 1200.$$

The core of this game is empty, since the standalone core conditions imply that the cost share of the first city, x_1, must satisfy

$$c^{\pi}(123) - c^{\pi}(23) = 400 \le x_1 \le 40 = c^{\pi}(1).$$

Another way to see this is to notice that we can split the full tour $0 \to 2 \to 1 \to 3 \to 0$ into two subtours: $0 \to 1 \to 0$ and $0 \to 2 \to 3 \to 0$. The full tour has cost 1200, but adding up the costs of the subtours (here $40 + 800 = 840$) will be cheaper.

The Shapley value $\phi^{Sh}(N, c^{\pi}) = (146.66, 706.66, 346.66)$ is violating individual rationality, since $x_1^{Sh} = 146.66 > c^{\pi}(1) = 40$.

Next, note that if we had chosen any of the orders,

$$\pi \in \{(123), (132), (231), (321)\},$$

where the total cost is minimized, the core of the associated fixed-route cost-allocation game is nonempty. Indeed, the core consists of allocations x, where $x_1 + x_2 + x_3 = 800$, and $0 \le x_1 \le 40$, $360 \le x_2 \le 800$, $0 \le x_3 \le 440$. This is actually a general result when costs satisfy the triangular inequalities (see section 3.1.3).

Finally, notice that it is *not* possible to define a uniquely determined irreducible cost matrix when the optimal graph is a cycle: in contrast to the MCST model from chapter 2.

3.1.3 Proportional Allocation

Consider a fixed route π. When the route is fixed, we write $C = C^{\pi}$ for the total cost of the tour, and $(N, c) = (N, c^{\pi})$ for associated cost-allocation problem.

Fishburn and Pollak (1983) propose to focus on the simple proportional rule with respect to the standalone cost $c(i)$ for each sponsor city $i \in N$:

$$\phi_i^P(N, c) = \frac{c(i)}{\sum_{j \in N} c(j)} C. \tag{3.3}$$

Since $C \le \sum_{j \in N} 2c_{0j} = \sum_{j \in N} c(j)$, we have that $\phi_i^P(N, c) \le c(i)$, for all $i \in N$. Thus, the proportional rule (3.3) satisfies individual rationality, which is arguably a desirable feature of any relevant allocation rule in this case.

Now consider the following additional requirements of any budget-balanced cost-allocation rule ϕ.

Cost monotonicity   Consider two cost structures $\{k_{ij}\}$ and $\{c_{ij}\}$, where $k_{0i} = c_{0i}$ for all $i \in N$, and $k_{ij} \geq c_{ij}$ otherwise. Then $\phi(N, k) \geq \phi(N, c)$.

In words, cost monotonicity (in the present version) states that if the costs between sponsor cities increase, then no sponsor can get a lower cost share.

Cost additivity   Consider two cost structures $\{k_{ij}\}$ and $\{c_{ij}\}$, where $k_{0i} = c_{0i}$ for all $i \in N$. Then $\phi(N, k + c) = \phi(N, k) + \phi(N, c)$.

In words, adding up the cost shares of two different cost structures (with identical individual round-trip costs) corresponds to sharing the costs of the aggregated problem.

Incremental solution   If $C = \sum_{i \in N} c_{0i}$, then $\phi_i(N, c) = c_{0i}$.

In words, if the total cost equals the sum of single-city trip costs, then each city should be allocated exactly this cost. For example, suppose that $c_{i,i+1} = c_{0,i+1}$ for $i = 1, \ldots, n$, with $c_{0n} = 0$. Then the total cost of the tour is $\sum_{j \in N} c_{0j}$, and it seems reasonable to assign cost shares to each sponsor city i that equal their incremental cost c_{0i}.

Fishburn and Pollak (1983) show that if all these requirements must be satisfied, in addition to individual rationality, only the proportional rule is relevant, as stated in theorem 3.1.

Theorem 3.1   Consider the fixed-route cost-allocation problem. A rule ϕ satisfies individual rationality, cost monotonicity, cost additivity, and incremental solution if and only if ϕ is the proportional rule given by (3.3).

Sketch of argument: It is clear that ϕ^P as defined in (3.3) satisfies all the axioms. Hence, consider the converse claim. Fishburn and Pollak (1983) show that if ϕ satisfies individual rationality, cost monotonicity, and additivity,

$$\phi_i = 2c_{0i} - \left(\sum_{j=1}^{n} 2c_{0j} - C \right) F_i(\{c_{0j}\}_{j \in N}),$$

where $\sum_{i \in N} F_i(\{c_{0j}\}_{j \in N}) = 1$. By incremental solution, we get $F_i(\{c_{0j}\}_{j \in N}) = \frac{c_{0i}}{\sum_{j \in N} c_{0j}}$, and consequently, $\phi_i = \phi_i^P$.

Remark   Notice that both cost monotonicity and additivity are restricted by the requirement $k_{i0} = c_{i0}$ for all $i \in N$. To demonstrate why the proportional rule does not satisfy cost monotonicity and additivity in general, let $k = (k_{10}, k_{20}, k_{12}) = (1, 1, 2)$, and $c = (c_{10}, c_{20}, c_{12}) = (2, 1, 2)$ (so $k_{10} \neq c_{10}$, and $c \geq k$), yielding $\phi^P(N, k) = (2, 2)$, and $\phi^P(N, c) = (3.33, 1.66)$. This contradicts cost monotonicity in general, since agent 2 pays less under c than under k. Moreover, $\phi^P(N, k+c) = (5.4, 3.6)$, contradicting cost additivity in general, since $\phi^P(N, k) + \phi^P(N, c) = (5.33, 3.66)$.

The proportional rule satisfies individual rationality. However, it is easy to show that the proportional rule does not necessarily satisfy all core conditions when the core is nonempty. Thus, when the tour between the n sponsor cities is cost efficient, there may still be good reasons to search for an allocation in the core.

As mentioned above, due to the triangular cost inequalities (3.1), we have that $C \leq \sum_{i \in N} 2c_{0i}$. Moreover, budget balance and individual rationality together imply that if $C = \sum_{i \in N} 2c_{0i}$, then $\phi_i = 2c_{0i}$. Hence the way that costs are allocated between edges ij ($i, j \in N$) on the tour is irrelevant for the solution in this case. This hints that the solution can only depend on the total cost of the tour as well as on all the one-way costs for each sponsor city c_{0i}.

Consequently, the problem closely resembles a rationing problem (see, e.g., Hougaard, 2009), where each sponsor city has a "claim" of its round trip cost $2c_{0i}$, and the total cost $C \leq \sum_{i \in N} 2c_{0i}$ has to be shared such that $0 \leq x_i \leq 2c_{0i}$ for each sponsor city. Parallel results to theorem 3.1 above can be found in Chun (1988, theorem 5) and Moulin (1987, theorem 5).

3.1.4 Limited Cost Information

Suppose that cost is known only for edges in the actually chosen tour involving all n nodes (agents), for example, if the cycle is an existing network and the cost of all other connections are unknown. So now there is no way to assess the standalone costs of each agent (or coalitions of agents, for that matter), and consequently, we cannot establish a game as in (3.2). Moreover, the proportional rule in the sense of (3.3) cannot be applied either (since standalone costs c_{0i} are unknown).

However, there is at least one straightforward way to extend the idea of proportional fairness to this new scenario. In a cycle all agents have two edge-disjoint paths to the root 0. When $c_{ij} = c_{ji}$, it seems natural to consider the minimum of these two path costs as the standalone option for each agent. Let the order of the cycle be given by $\pi = 1, 2, \ldots, n$ (otherwise, relabel the agents) and let

$$\tilde{c}(i) = \min \left\{ \sum_{j=0}^{i-1} c_{j(j+1)}, \sum_{j=i}^{n-1} c_{j(j+1)} + c_{n0} \right\}$$

be the relevant standalone cost of agent i. Then total costs C^{π} can be allocated according to proportional shares:

$$\phi_i^P = \frac{\tilde{c}(i)}{\sum_{j=1}^n \tilde{c}(j)} C^{\pi}. \tag{3.4}$$

Note that

$$\phi_i^P = \frac{\tilde{c}(i)}{\sum_{j=1}^n \tilde{c}(j)} C^{\pi} = \tilde{c}(i) + \frac{\tilde{c}(i)}{\sum_{j=1}^n \tilde{c}(j)} \left(C^{\pi} - \sum_{j=1}^n \tilde{c}(j) \right).$$

Thus, we can interpret $\tilde{c}(i)$ as the minimum cost that each agent i has to cover, and on top of that, the proportional rule makes i pay a proportional share of the excess cost of the entire tour $\left(C^{\pi} - \sum_{j=1}^n \tilde{c}(j)\right)$.

The proportional rule (3.4) has a desirable property with respect to the number of agents that a single node in the cycle represents. Suppose that a node on the cycle can be a common connection point of several agents with individual links to that node, as illustrated in the figure below (where nodes a and b only represent one agent, while node c represents two agents, i' and i'').

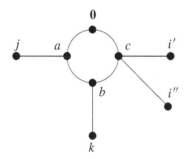

Sharing the total cost of edges in the loop, the proportional rule assigns the same cost share to the common connection point, no matter how many agents it claims to represent. Indeed, suppose that i represents two agents, i' and i'', such that $\tilde{c}(i) = \tilde{c}(i') + \tilde{c}(i'')$; then we have $\phi_i^P = \phi_{i'}^P + \phi_{i''}^P$. In comparison, for instance, the equal-split rule can be manipulated by claiming that a node only represents a single agent, when in fact it represents multiple agents. Such concerns may be relevant in server networks or electrical grids, where agents hook up to a source loop with individual connections.

3.1.5 Limited Budgets

Fishburn (1990) considers a generalization of the model, in which sponsor cities have limited budgets, with city i's (maximal) willingness to pay being w_i, for $i \in N$. Note that in the model above, $w_i = 2c_{0i}$ for all i, but there may be cases where some sponsors are willing to pay less than a round trip cost to their city, while others are willing to pay more and thereby make the entire tour feasible. Let $W = \sum_{i \in N} w_i$ be the total willingness to pay, and denote by $F(c) = \{\pi \in \Pi \mid c^\pi(N) \leq W\}$ the set of feasible tours.

Moreover, assume that costs satisfy the triangular inequalities in general; that is, for any $i, j, k \in N^0$,

$$c_{ik} + c_{jk} \geq c_{ij}, \tag{3.5}$$

which is obviously a stronger condition than (3.1).

Now consider a budget-balanced cost-allocation rule ϕ for feasible tours $\pi \in F(c)$, and the following new requirements.

Independence of nonconsecutive edge costs For $\pi \in F(c)$, the cost shares $\phi_i(N, c^\pi)$ are independent of costs c_{jk}, where $j, k \notin \{\pi(l), \pi(l+1)\}: 1 \leq l \leq n-1$.

In words, cost shares do not depend on costs between sponsor cities j and k that are not consecutive in the tour π. Note that this property is satisfied by the proportional rule (3.3), since cost shares can still depend on the round trip costs to the home city 0.

The next axiom states that the cost share of a sponsor is bounded from above by the sponsor's maximum willingness to pay.

Budget upperbound For $\pi \in F(c)$, $\phi_i(N, c^\pi) \leq w_i$ for all $i \in N$.

The final new property relates the cost of two different feasible tours, stating that if one tour π is more expensive than another tour π', then no sponsor can gain by insisting on tour π instead of π'. In this way it is a kind of solidarity requirement: if a more expensive tour is chosen, every sponsor pays at least the same as before. This property is formalized as follows.

Tour cost monotonicity For $\pi, \pi' \in F(c)$, if $c^\pi(N) > c^{\pi'}(N)$, then $\phi(N, c^\pi) \geq \phi(N, c^{\pi'})$.

Together with the axioms of cost monotonicity and additivity from before, Fishburn (1990) is now able to show the following result.

Theorem 3.2 A (budget-balanced) cost-allocation rule satisfies independence of nonconsecutive cost, budget upper bound, cost monotonicity, tour cost monotonicity, and cost additivity if and only if there are weights $\lambda_i \geq 0$ with

$\sum_{i \in N} \lambda_i = 1$, and $a_i \in \mathbb{R}$ with $\sum_{i \in N} a_i = 0$ such that, for all $i \in N$ and feasible tours $\pi \in F(c)$,

$$\phi_i(N, c^\pi) = a_i + \lambda_i c^\pi(N).$$

Moreover, if $W \leq \sum_{i \in N} 2c_{0i}$, then $a_i = w_i - \lambda_i W$, and if $W \geq \sum_{i \in N} 2c_{0i}$, then $a_i \leq w_i - \lambda_i \sum_{i \in N} 2c_{0i}$.

No particular rule is singled out by the theorem. In fact, several rules present themselves rather naturally. Indeed, note that in case $c^\pi(N) = W$, then $\phi_i(N, c^\pi) = w_i$ for all $i \in N$ (implied by budget balance and budget upper bound). Thus, a natural rule to consider is the proportional rule with respect to willingness to pay w_i,

$$\phi_i^P(N, c^\pi) = \frac{w_i}{W} c^\pi(N),$$

for all $i \in N$. Clearly, this rule corresponds to setting $\lambda_i = w_i/W$, which implies that $a_i = 0$ for all i. But we may, for example, also consider the rule where $\lambda_i = \frac{1}{n}$ such that the cost shares are determined by subtracting the same amount $\frac{W}{n} - \frac{c^\pi(N)}{n}$ from each individual willingness to pay w_i, $\phi_i(N, c^\pi) = w_i - \frac{1}{n}(W - c^\pi(N))$, for all $i \in N$ Note that $W - c^\pi(N) \geq 0$ for all $\pi \in F(c)$.

Yet another allocation rule that may have some appeal in this case is the constrained equal-split rule, where the total cost of a feasible tour C is split equally, provided that no sponsor pays more than w_i, that is,

$$x_i = \min\{w_i, \alpha\},$$

where α is chosen such that $\sum_{i \in N} x_i = C$. However, this rule does not satisfy cost additivity. Indeed, take the two-sponsor case where $c = (c_{01}, c_{02}, c_{12}) = (5, 6, 1)$, and $k = (k_{01}, k_{02}, k_{12}) = (5, 6, 11)$, and let $w = (10, 12)$. Clearly, c and k satisfy the triangular inequalities (3.5). Looking at the (feasible) tour $\pi = 0120$ for c, the total cost 12 can be split equally without violating individual budgets: $x(c) = (6, 6)$. Looking at π for k, the total cost 22 yields a constrained equal split $x(k) = (10, 12)$. However, $c + k = (10, 12, 12)$ with $w' = (20, 24)$, yielding $x(c + k) = (17, 17) \neq x(c) + x(k) = (16, 18)$.

3.2 Traveling Salesman

Example 3.1 showed that there is no guarantee of a nonempty core in the fixed-route cost-allocation game (3.2), simply because the total cost may not be minimized among all the options of a tour between the n sponsor cities and the home city. In fact, if costs satisfy the triangular inequalities (3.1), and the total cost given the fixed order $\bar{\pi}$ is minimized (i.e, $c^{\bar{\pi}}(N) = \min_{\pi \in \Pi} \{c^{\pi}(N)\}$), then the core of (3.2) is nonempty (Potters et al., 1992). Indeed (following the argument in Derks and Kuipers, 1997), consider an arbitrary $S \subseteq N$, and let $\bar{\pi}$ be a cost-minimal tour. Let $\bar{\pi}_1$ be the tour resulting from skipping the agents in $N \setminus S$, and let $\bar{\pi}_2$ be the tour resulting from skipping the agents in S. Then construct a new tour π as follows. Start in 0, and follow the tour in $\bar{\pi}_1$ up to the last agent in S, say, agent i. Then jump to the first agent in the tour $\bar{\pi}_2$, say, agent j, and follow the tour $\bar{\pi}_2$ to 0. By (3.1), we have $c_{0i}^{\pi} + c_{0j}^{\pi} \geq c_{ij}^{\pi}$, so $C^{\pi} \leq C^{\bar{\pi}_1} + C^{\bar{\pi}_2}$. Thus, since $\bar{\pi}$ is minimal, we have $C^{\bar{\pi}} \leq C^{\pi} \leq C^{\bar{\pi}_1} + C^{\bar{\pi}_2}$; that is, $c(N) \leq c(S) + c(N \setminus S)$ (which is equivalent to the game having a nonempty core).

In other words, if costs satisfy the triangular inequalities (3.1), then selecting a cost-minimizing tour guarantees the existence of stable cost allocations in the fixed-route cost-allocation game.

Yet, if we allow the grand coalition N to minimize their tour cost, why not also let every subset of sponsor cities be allowed to minimize the total cost of their subtour as well, and use this as the relevant standalone cost of each coalition? Indeed, this is the idea of the traveling salesman game.

3.2.1 Traveling Salesman Game

For any $S \subseteq N$, let π^S be a cyclic tour among cities $S^0 = S \cup \{0\}$ (starting and ending in the home city 0, where each sponsor city in S is visited exactly once), and let $c^{\pi^S}(S) = c_{01} + c_{12} + \cdots + c_{(s-1)s} + c_{0s}$ be the total cost of the (ordered) tour π^S. Now, for all $S \subseteq N$, let

$$\bar{c}(S) = \min_{\pi^S} \{c^{\pi^S}(S)\}, \tag{3.6}$$

and let

$$(N, \bar{c}), \tag{3.7}$$

with $\bar{c}$ as defined in (3.6) denote a *traveling salesman game*. By (3.6), each coalition of sponsor cities is allowed to find a cost-minimizing tour involving the home city. In other words, in the traveling salesman game, a minimum-weight Hamiltonian cycle is found for each (complete) subgraph involving the nodes in S^0. If weights (costs) satisfy the triangular inequalities (3.5) (and the graph is complete, as in our case) then such a minimum-weight Hamiltonian cycle exists for each subgraph for which no other cycle (even cycles that may visit a sponsor city more than once) has lesser weight.[1] Algorithms for finding

---

1. For arbitrary graphs it is clear that there may not exist a Hamiltonian cycle for each subgraph, and even if they do exist, there may be another cycle that visits sponsor cities more than once that has a smaller weight.

a minimum-cost Hamiltonian cycle for each coalition in the traveling salesman game can be found, for example, in Jünger et al. (1995).

Example 3.2 (A class of problems with a nonempty core)   Consider the class of traveling salesman problems where all edge costs are identical (say, normalized to 1). Clearly, $\bar{c}(S) = s + 1$ for all coalitions $S \subseteq N$. Since $s(n+1)/n \leq s+1$, for $s = 1, \ldots, n$, the equal-split allocation $\left(\frac{n+1}{n}, \ldots, \frac{n+1}{n}\right)$ satisfies the core conditions.

Given costs that satisfy the triangular inequalities (3.5), the core of the traveling salesman game (3.7) is nonempty for up to five sponsor cities (while examples of games with empty cores can be found for $n \geq 6$); see Tamir (1989) and Kuipers (1993). Potters et al. (1992) provide an example of a traveling salesman game with an empty core and four sponsor cities, where costs are asymmetric, that is, $c_{ij} \neq c_{ji}$. They further provide sufficient conditions on edge costs for a nonempty core for (3.7).

In particular, it is shown in Curiel (1997) that if the sponsor cities are clustered far away from the home city, then the cost savings from cooperation are sufficiently large to ensure a nonempty core. To be more precise: if, for all $i, j \in N$,

$$\frac{c_{0i} + c_{0j}}{n} \geq \max_{\pi \in \Pi}\{c_{\pi(1)\pi(2)} + c_{\pi(2)\pi(3)} + \cdots + c_{\pi(n-1)\pi(n)}\},$$

then the core of (3.7) is nonempty.

So allowing every coalition (and not just the grand coalition consisting of all sponsor cities), to minimize tour costs, introduces the possibility of an empty core if there are more than five sponsor cities. When the core is empty, we can of course still use the proportional rule, but the question is whether this is more reasonable than looking at the fixed-route game for a cost-minimizing cycle and choosing an allocation in the core. Clearly, such considerations will depend sensitively on the particular application at hand, and the decisions are therefore somewhat ad hoc.

3.2.2   Practical Application

In practice, agents may also differ in size, or demand, which may be relevant when sharing costs, depending on the specific application. Engevall et al. (1998) present an example of a practical application related to distribution of gasoline to gas stations by the company Norsk Hydro Olje AB in the southern part of Sweden. The gas stations have different demands, and costs are given as transportation costs. The depot is the home city, and the gas stations are the sponsor cities. A truck driving from the depot needs to find the cost-minimizing tour that supplies gasoline to all the gas stations. With suitable simplifications, this can be formulated as a traveling salesman game, since transportation costs are symmetric and satisfy the triangular inequalities. In an example involving five gas stations, it is shown how the result of using the various game-theoretic solution concepts differs from the actual cost allocation by Norsk Hydro. It is also shown how demand differences between the gas stations can be taken into account.

In addition, several trucks may be used due to capacity or time constraints, as in the famous vehicle routing problem (see, e.g., Golden et al. 2008). Here, the optimal solution involves several tours (one for each truck with a given capacity and time window), and it becomes much more complex to define overall fairness requirements when transportation costs must be allocated among the customers. It is far from obvious how customer demands for pick-up or delivery times, as well as the size of their demands, should influence allocation issues due to the complex interplay of these data in the final planning of the optimal route. Because of the size of most optimization problems in practice, we are in effect prevented from using the game-theoretic approach by simple computational constraints. Thus, the challenge of finding compelling allocation rules seems primarily related to the specific case at hand, and it remains an open question whether the modern heuristics used to solve the optimization problems can form the basis for reasonable notions of fair allocation.

3.3 Chinese Postman

A third variant of routing games is called the *Chinese postman game* (see, e.g., Kwan, 1962; Hamers et al., 1999). The story goes as follows. A postman delivers mail to a set of agents (here edges in a connected graph). Each edge is associated with a cost. The postman starts and ends at the post office (the root), and he organizes the tour to minimize total costs, given that each edge has to be visited at least once. For an algorithm to determine the efficient (cost-minimizing) tour, see, for example, Edmonds and Johnson (1973).

Note that to minimize costs, the postman may visit some edges more than once. The natural game associated with the Chinese postman problem allocates the nonseparable costs, that is, the total (minimal) cost of the efficient tour minus the sum of edge costs for which each individual agent is directly liable.

3.3.1 The Model and the Game

Let $G = (N^0, E)$ be a connected undirected graph. Let $e_i \in E$ be the ith edge and c_i its associated cost. Let the set of agents be identified by the set of edges E, and let node $0 \in N^0$ be the root (post office) in G. For some subset $S \subseteq E$, an

S-tour in G (with respect to 0) is a cycle $(0, e_1, \ldots, e_k, 0)$ starting at 0, visiting each edge in S at least once, and then returning to 0. Note that an S-tour may include edges outside S. Let $T(S)$ denote the set of such S-tours in G.

For each subset $S \subseteq E$, let the nonseparable costs of the related S-tour be defined by

$$
c(S) = \min_{(0, e_1, \ldots, e_k, 0) \in T(S)} \left\{ \sum_{j=1}^{k} c_j - \sum_{j \in S} c_j \right\} \tag{3.8}
$$

The pair (E, c), where c is a cost function defined by (3.8), is called a *Chinese postman game*.

The following example (from Hamers, 1997) demonstrates that the Chinese postman game may have an empty core.

Example 3.3 (Chinese postman game with empty core) Consider the connected graph below with root 0, and five edges (agents) $e_1, \ldots, e_5$.

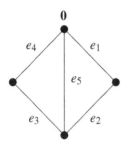

Let the costs be identical and normalized to $c_i = 1$ for all $i \in E$. Thus, by (3.8) we have that the total (nonseparable) cost to be divided among the agents (edges) is $c(N) = 6 - 5 = 1$, since a cost-minimizing tour (for instance, the tour $(0, e_1, e_2, e_3, e_4, e_5, e_5, 0)$) is visiting e_5 twice. Since $c(S) = 0$ for $S \in \mathcal{S} = \{(1, 2, 5), (3, 4, 5), (1, 2, 3, 4)\}$, we have that

$$
2 = 2c(N) = \sum_{S \in \mathcal{S}} \sum_{i \in S} x_i \leq \sum_{S \in \mathcal{S}} c(S) = 0,
$$

violating the core conditions. The Shapley value is given by the payment vector $(x_1, x_2, x_3, x_4, x_5) = (0.07, 0.32, 0.32, 0.07, 0.22)$, making agents (edges) 2 and 3 cover the majority of the (nonseparable) cost, which may seem somewhat odd, given the symmetric nature of the graph.

Before we can characterize the class of Chinese postman games with nonempty cores, we need a few definitions. Recall that an edge $e \in G$ is called a *bridge* if $G - e$ is disconnected. Let $B(G)$ be the set of bridges in G. A graph

G is *weakly Eulerian* if, after removing all bridges $e \in B(G)$, the remaining components are all Eulerian, that is, the degree of every node is even (see chapter 1), or singleton.

It is clear that if the graph is weakly Eulerian, the associated Chinese postman game has a nonempty core, since the allocation, $x_i = c_i$ if $e_i \in B(G)$ and $x_i = 0$ otherwise, satisfies the standalone core conditions (indeed: $\sum_{i \in S} x_i = \sum_{b \in B(G) \cap S} c_b \leq c(S)$ for arbitrary $S \subseteq N$). In fact, G being weakly Eulerian is both a necessary and sufficient condition for a nonempty core of the Chinese postman game, as shown in Granot et al. (1999).

Using a seven-node complete graph (which, since all nodes have degree 6, is Eulerian and so weakly Eulerian), Hamers (1997) demonstrates that the Chinese postman game need not be concave. Yet if the graph is weakly cyclic (i.e., if all remaining components of G after removing all bridges are singletons, circuits, or unions of circuits, where each pair of circuits in the union has at most one node in common), then the Chinese postman game is concave. In fact, the graph being weakly cyclic is both a necessary and sufficient condition for concavity of the Chinese postman game, as shown in Granot et al. (1999).

3.3.2 A Particular Allocation Rule

The fact that existence of stable solutions (in the sense of the core) is tantamount to an underlying weakly Eulerian graph can be used to suggest a specific allocation rule that differs from the usual game-theoretic solution concepts by being directly related to the structure of the network.

Consider the class of Chinese postman games arising from weakly Eulerian graphs, denoted $\mathcal{WE}$. For a given weakly Eulerian graph G and bridge $b \in B(G)$, an edge $e \in E$ is called a *follower* of b with respect to the root 0 if each path that contains both 0 and e also contains b. Let $F_b(G)$ denote the set of followers of b in G (and note that $b \in F_b(G)$). Since the postman needs to pass the bridge b twice when he delivers mail to members in $F_b(G)$, it seems fair that all members of $F_b(G)$ share the added (nonseparable) cost of c_b equally. Noting that several bridges may be passed in order to reach a given edge, this edge should pay its fair share of every bridge passed. Formally, for problems $(E, c) \in \mathcal{WE}$, we define the allocation rule ϕ^E by cost shares, for every edge $e \in E$, as

$$\phi_e^E (E, c) = \sum_{b \in B(G): e \in F_b(G)} \frac{c_b}{|F_b(G)|}. \tag{3.9}$$

Clearly, this rule is budget balanced, since $c(E) = \sum_{b \in B(G)} c_b = \sum_{e \in E} \phi_e^E$. On the class $\mathcal{WE}$, the rule ϕ^E results in core allocations.[2]

---

2. See exercise 3.9.

In many ways the idea behind the allocation rule ϕ^E appears to resemble the serial principle (or equal share downstream idea) for tree-structured graphs, but when applied to cycles, we lose the close relation to the Shapley value, as demonstrated by the following example.

Example 3.4 (Comparing ϕ^E to the Shapley value) Consider the two weakly Eulerian 4-edge graphs, G and G' below, where only the position of the root 0 differs. Clearly e_4 is a bridge connecting a singleton node with the circuit (e_1, e_2, e_3). So in fact, the graph is weakly cyclic, and the associated Chinese postman game is therefore concave. Let edge costs be normalized to $c_e = 1$ for all $e \in E$.

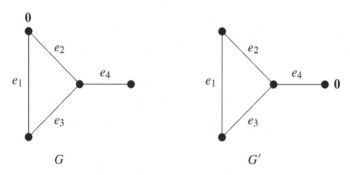

We have that $F_{e_4}(G) = \{e_4\}$ and $F_{e_4}(G') = \{e_1, e_2, e_3, e_4\}$, so using the allocation rule (3.9), we have

$$\phi^E(E, c^G) = (0, 0, 0, 1) \quad \text{and} \quad \phi^E(E, c^{G'}) = (0.25, 0.25, 0.25, 0.25).$$

The associated Chinese postman games are given in the table below

S	$c^G(S)$	$c^{G'}(S)$
e_1	1	4
e_2	1	3
e_3	2	3
e_4	3	1
e_1, e_2	1	3
e_1, e_3	1	3
e_1, e_4	3	3
e_2, e_3	1	3
e_2, e_4	2	2
e_3, e_4	3	2

S	$c^G(S)$	$c^{G'}(S)$
e_1, e_2, e_3	0	2
e_1, e_2, e_4	2	2
e_1, e_3, e_4	2	2
e_2, e_3, e_4	2	2
e_1, e_2, e_3, e_4	1	1

Both allocations $\phi^E(E, c^G)$ and $\phi^E(E, c^{G'})$ are obviously core allocations in their respective games. Compared to the Shapley values (which are also core allocations here, since both games are concave) given by $\phi^{Sh}(E, c^G) = (-0.25, -0.42, 0.08, 1.58)$ and $\phi^{Sh}(E, c^{G'}) = (0.83, 0.33, 0.33, -0.50)$ there is a clear difference. In particular, the Shapley values involve transfers between the agents (as indicated by negative payments).

Hamers et al. (1999) show that the allocation rule (3.9) can be characterized by "bridge-cluster symmetry," stating that each group of agents (edges) that needs the same set of bridges to be connected to the root should contribute equally to the nonseparable costs, and an independence property, stating that agents not in a bridge cluster containing some extreme bridge (i.e., a bridge with no other bridge as follower) should pay the same, independently of whether this bridge cluster is organized as in the original graph or reorganized as a circuit now containing the extreme bridge in question.

3.4 k-Connectivity and Reliability

In routing problems, cycles arise directly as the result of the modeled activity, but cycles may also be part of a deliberate network design in many other cases. For instance, when networks function as distribution systems, connections (edges) may fail to deliver or may be temporarily blocked. Having access to alternative paths to the supplier (root) can therefore be valuable for network users as a buffer against service disruption. Such reliability concerns can be addressed when designing the optimal network structure simply by allowing agents to have multiple distinct paths to the source, although strictly speaking, they are not necessary to obtain the desired source connection.

Cycles may also be the result of a historical evolution of the network, where, for instance, old connections gradually become replaced by new ones, and it is too costly to destroy or dismantle those that are obsolete. For such reasons an existing network, at a given point in time, may contain cycles that include redundant connections, but they are operational and provide some users with improved connectivity.

Power grids and other types of electrical transmission structures are good examples of networks that may include cycles, for example, if constructed as *source loops*, where consumers (relevant nodes) are connected to the source (e.g., a power plant or transmission station) via an individual connection to a loop (cycle) on which the source node is located. In this way consumers are ensured increased reliability of the network, since they have at least two distinct paths to the source via the loop. Thus, if there is failure (or maintenance) on some connections in the loop, consumers can typically still be served via the remaining operational connections.

In the context of cost allocation by a central planner, we will now consider fixed (existing) networks involving a source loop. Since the network is fixed, we do not know the cost of alternative network options, and the existing network structure may even include redundant connections (for instance, if the network provide users with higher connectivity than they require). Consequently, a game-theoretic approach to solve the allocation problem is less natural, since standalone costs are counterfactual. And even if we insist on defining some type of standalone cost, the presence of redundant connections typically results in games with an empty core.

3.4.1 k-Connectivity

For simplicity, consider an existing source-loop structure represented by a graph $G = (N^0, E)$, where there is one source node 0 (the supplier) located on the loop, and each of the remaining nodes $i \in N$ represent an agent demanding connection to the source. Edges are public goods (i.e., can be used by everyone without rivalry) and are costly to operate or maintain. Let $c_e > 0$ be the cost associated with edge $e \in E$.

In any type of rooted graph, we say that a node $i \in N$ has k-*connectivity* if there are k pairwise distinct paths from node i to the source 0. We use the term *strong* k-connectivity if these k paths are mutually edge disjoint. Source loops (as mentioned above) offer at least 2-connectivity to all agents and are therefore more reliable than spanning trees, which offer only 1-connectivity to everyone. For all $i \in N$, let $c_i(G)$ denote the connectivity of node i in graph G (either in weak or strong form).

Example 3.5 (Computing connectivity)   Consider the source loop G below, where three agents, Ann, Bob, and Carl, are connected to the loop through individual connections Aa, Bb, and, Cc, respectively, and the source 0 is located on the loop itself.

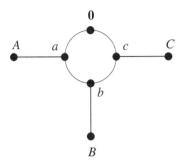

Clearly all three agents have 2-connectivity, since they all have two distinct paths to the source via the loop;

Ann :$\{Aa, a0\}, \{Aa, ab, bc, c0\},$

Bob :$\{Bb, ab, a0\}, \{Bb, bc, c0\},$

Carl :$\{Cc, c0\}, \{Cc, bc, ab, a0\},$

Thus, $c_i(G) = 2$ for all three agents.

If we add the link ac to the source loop G, we obtain the new graph $G' = G + ac$, as illustrated below (for example, ac may be an old connection that is too costly to dismantle).

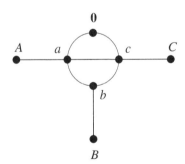

This added connection changes the connectivity of the agents. The following possible distinct paths are now added to those of the original graph:

Ann :$\{Aa, ac, c0\},$

Bob :$\{Bb, ab, ac, c0\}, \{Bb, bc, ac, a0\},$

Carl :$\{Cc, ac, a0\},$

yielding connectivities, $c_{Ann}(G') = c_{Carl}(G') = 3$, and $c_{Bob}(G') = 4$. This indicates that the edge ac is more valuable to Bob than to Ann and Carl in terms of connectivity.

In the context of cost allocation, the connectivity index can be used to indicate agents' private value of the network. In the absence of other forms of relevant preference information, a straightforward approach is therefore to allocate total cost in proportion to connectivity. In particular, when edge-specific costs are available, a similar straightforward form of approach is to allocate the cost of each edge in proportion to an edge-specific index of *marginal connectivity*, either in absolute or relative terms.

Formally, for a given graph G, define the *absolute marginal connectivity* of edge $e \in G$, for node $i \in N$ as

$$\theta_i^c(G, e) = c_i(G) - c_i(G - e). \tag{3.10}$$

That is, the absolute marginal connectivity is the absolute difference in agent i's connectivity with, and without, the edge e.

Note that $\theta_i^c(G, e) = 0$ when e is *irrelevant* for agent i (i.e., when e is not part of any of i's possible paths to the source), and $\theta_i^c(G, e) = c_i(G)$ when e is *critical* to agent i (i.e., when e is part of every possible path to the source). When i has strong k-connectivity ($k \geq 2$), no edge is critical.

Likewise, define the *relative marginal connectivity* of edge $e \in G$, for node $i \in N$, as

$$\theta_i^{rc}(G, e) = \frac{c_i(G) - c_i(G - e)}{c_i(G)}. \tag{3.11}$$

Clearly, $\theta_i^{rc}(G, e) = 0$ when e is irrelevant for agent i, and $\theta_i^{rc}(G, e) = 1$ when e is critical to agent i.

For every edge e, the cost c_e can now be allocated in proportion to marginal connectivity. That is, the cost share for agent i is given by

$$x_i(G, c_e) = \frac{\theta_i(G, e)}{\sum_{j \in N} \theta_j(G, e)} c_e \tag{3.12}$$

for all $i \in N$ and $e \in G$, where θ_i may be given by (3.10) or (3.11) respectively. Denote by

$$\varphi_i(G, e) = \frac{\theta_i(G, e)}{\sum_{j \in N} \theta_j(G, e)} \in [0, 1]$$

the *cost ratio* of agent i based on the absolute or relative marginal connectivity measure (in general, though, θ could be any function describing i's *liability* associated with the edge e; see Hougaard and Moulin 2014, and section 4.2).

In chapter 4 we return to cost-allocation rules where the cost of each edge is shared separately according to cost ratios, and final cost shares are added up over the set of edges:

$$\phi_i(G, c) = \sum_{e \in G} x_i(G, c_e) = \sum_{e \in G} c_e \varphi_i(G, e), \quad \text{for all } i \in N. \tag{3.13}$$

Chapter 4 provides a more general and deeper analysis. In the computer science literature, such separable rules are often called *protocols*, (see, e.g., Chen et al. (2010), and chapter 5). Protocols are intimately connected with the property of cost additivity. Indeed, if a rule ϕ satisfies cost additivity (i.e., $\phi(G, c' + c'') = \phi(G, c') + \phi(G, c'')$), then it is a protocol and can be written as (3.13).

Example 3.5 (continued: Comparing absolute and relative marginal connectivity indices) Looking at the graph G in example 3.5 above, it seems rather natural that Ann, Bob and Carl should pay for their own connection to the loop Aa, Bb, and Cc, respectively. In contrast, the cost of the loop itself seems naturally to be shared equally. Indeed, this is also the result of using proportional sharing with respect to both absolute and relative marginal connectivity, since

$$\theta^c_{Ann}(G, Aa) = \theta^c_{Bob}(G, Bb) = \theta^c_{Carl}(G, Cc) = 2,$$

and $\theta^c_i(G, Jj) = 0$ for all $i \neq j$. Moreover, $\theta^c(G, e) = 1$ for all other (loop) edges and agents. So the cost of individual connections to the loop will be shared using ratios $(1, 0, 0)$, $(0, 1, 0)$, and $(0, 0, 1)$ respectively, while all loop edges will be shared according to the ratios $(1/3, 1/3, 1/3)$.

More interestingly, we can consider the redundant link ac in the graph G' in example 3.5 above. Here, we have that $\theta^c_{Ann}(G', ac) = \theta^c_{Carl}(G', ac) = 1$, and $\theta^c_{Bob}(G', ac) = 2$. Therefore, the cost c_{ac} will be shared using the ratios $(1/4, 1/2, 1/4)$. Relative marginal connectivity is slightly different, since $\theta^{rc}_{Ann}(G', ac) = \theta^{rc}_{Carl}(G', ac) = 1/3$, and $\theta^{rc}_{Bob}(G', ac) = 2/4 = 1/2$. Thus, the cost c_{ac} will be shared according to the ratios $(2/7, 3/7, 2/7)$ when we use θ^{rc} instead.

Adding the edge ac also changes the marginal connectivity of the other edges in the loop, and so also changes the agents' liability when allocating the cost of these loop edges. Results on the marginal connectivity of the four loop edges are presented in the table below.

	$a0$	$c0$	ab	bc
θ^c_A	1	2	1	1
θ^c_B	2	2	2	2
θ^c_C	2	1	1	1

	$a0$	$c0$	ab	bc
θ_A^{rc}	1/3	2/3	1/3	1/3
θ_B^{rc}	1/4	2/4	1/4	1/4
θ_C^{rc}	1/3	2/3	1/3	1/3

Clearly, the cost of these loop edges is no longer shared equally if allocated according to (3.12).

3.4.2 Reliability

The issue of network reliability may seem more appropriately modeled in a probabilistic framework, since basically we are trying to assess the probability of successfully connecting a given agent to the source when edges may fail at random. So network reliability for agent i can be measured as the probability that i can connect successfully to the source, taking the random failure of every edge directly into account. Similar to the index of connectivity, network reliability can be seen as a proxy for network value in the absence of other forms of relevant information about network users' preferences.

For simplicity, assume that each edge e in the existing network G has an exogenously determined probability ε of failure, and that these probabilities are stochastically independent and identical for all edges.

Denote by $\mathcal{D}^i \subseteq \mathcal{P}(E)$ the set of all subsets of edges in G that allow agent i to connect to the source (i's service sets in G). The sets in $\mathcal{D}^i$ are naturally inclusion-monotonic in the sense that if i gets source connection with the set of edges $S \in \mathcal{D}^i$, then i also gets source connection for any superset of S.

The probability that edges in S work, while edges in $E \setminus S$ fail, is $(1 - \varepsilon)^{|S|} \varepsilon^{|E \setminus S|}$. Thus, agent i's probability of successful source connection in the network G is given by

$$p_i(G, \varepsilon) = \sum_{S \in \mathcal{D}^i} (1 - \varepsilon)^{|S|} \varepsilon^{|E| - |S|}. \tag{3.14}$$

Example 3.6 (Computing reliability) Consider a source loop given by the graph G below, where the three agents Ann, Bob, and Carl, as well as the source 0, are located directly on the loop.

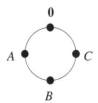

All agents have (strong) 2-connectivity:

Ann : $\{A0\}, \{AB, BC, C0\},$

Bob : $\{AB, A0\}, \{BC, C0\},$

$Carl$: $\{C0\}, \{BC, AB, A0\}.$

Now suppose that each of the four loop edges $\{A0, AB, BC, C0\}$ has a 5 percent chance of failure (i.e., $\varepsilon = 0.05$).

To compute Ann's probability of successful source connection, $p_A(G, 0.05)$, first note that

$$\mathcal{D}^A = \{(A0), (A0, AB), (A0, BC), (A0, 0C), (AB, BC, 0C)$$

$$(A0, AB, BC), (A0, AB, 0C), (A0, BC, 0C), (A0, AB, BC, 0C)\}.$$

Thus, using (3.14), we have

$$p_A(G, 0.05) = 0.95^4 + 4 \times 0.95^3 \times 0.05 + 3 \times 0.95^2 \times 0.05^2$$

$$+ 0.95 \times 0.05^3 = 0.9929,$$

since Ann will have access to the source if all four edges work, any (of the four possible) combinations of three edges work, and if the three possible combinations of two edges involving $A0$ work, as well as if only $A0$ works.

Clearly, Carl's reliability is the same as Ann's since they are completely symmetric in the graph, $p_C(G, 0.05) = p_A(G, 0.05) = 0.9929$. In contrast, Bob's reliability, $p_B(G, 0.05) = 0.9905$, is slightly lower—a result of the fact that both Ann and Carl have a direct source connection.

If, for example, the probability of failure is increased to 50% (i.e., $\varepsilon = 0.5$), we now have reliabilities $p_A(G, 0.5) = p_C(G, 0.5) = 0.5325$, and $p_B(G, 0.5) = 0.4375$.

Algorithms for computation of such (and related) probability measures can be found, for example, in Ball (1979) and Ball et al. (1995).

Since increased connectivity leads to higher probability for successful source connection, there is some degree of correlation between connectivity and reliability measures. But by using the reliability index, we are able to express that the size of the failure probability, ε, matters in the sense that users with identical connectivities may have different reliabilities (as illustrated by example 3.6 above, where all three agents have the same connectivity but with increasing ε, Bob's reliability becomes relatively lower compared to Ann's and Carl's).

When allocating total cost, we can therefore also choose to allocate in proportion to reliability, such that cost shares are determined by

$$\phi_i(G, c) = \frac{p_i(G, \varepsilon)}{\sum_{j \in N} p_j(G, \varepsilon)} \sum_{e \in G} c_e \qquad (3.15)$$

for all $i \in N$. This rule is dubbed the exante service rule in Hougaard and Moulin (2017), since it is mainly relevant when costs are shared (exante) before the resolution of uncertainties.[3]

As with measures of connectivity, we can also choose to allocate costs of separate edges using a measure of *marginal reliability* (both in absolute and relative terms) of a given edge, for a given agent in the graph G. Again the overall idea is that marginal reliability of the edge e is a proxy for agent i's value of e in terms of reliability: the difference in probability of successful source connection with, and without, the edge.

Define the *absolute marginal reliability* of edge $e \in G$, for node $i \in N$, as

$$\theta_i^p(G, e) = p_i(G, \varepsilon) - p_i(G - e, \varepsilon). \qquad (3.16)$$

Note that $\theta_i^p(G, \varepsilon) = 0$ when e is irrelevant for agent i, and $\theta_i^p(G, \varepsilon) = p_i(G, \varepsilon)$ when e is critical to agent i.

Likewise, define the *relative marginal reliability* of edge $e \in G$, for node $i \in N$ as

$$\theta_i^{rp}(G, e) = \frac{p_i(G, \varepsilon) - p_i(G - e, \varepsilon)}{p_i(G, \varepsilon)}. \qquad (3.17)$$

Note that $\theta_i^{rp}(G, \varepsilon) = 0$ when e is irrelevant for agent i, and $\theta_i^{rp}(G, \varepsilon) = 1$ when e is critical to agent i.

Clearly, both (3.16) and (3.17) can be used in the context of cost-allocation protocol (3.12).

Example 3.7 (Comparing marginal reliability indices) The absolute marginal reliability of the four edges in the source loop G from example 3.6 is given in the table below for $\varepsilon = 0.05$. The associated cost ratios φ_i are also listed. (Since the reliability is close to 1, there is almost no difference between absolute and relative marginal reliability in this case).

	$A0$	AB	BC	$C0$
$\theta_A^p(G, 0.05)$	0.1355	0.0439	0.0439	0.0439
$\theta_B^p(G, 0.05)$	0.0880	0.0880	0.0880	0.0880
$\theta_C^p(G, 0.05)$	0.0439	0.0439	0.0439	0.1355

---

3. Further discussion is postponed until section 4.2.5.

	$A0$	AB	BC	$C0$
φ_A	0.5067	0.2497	0.2497	0.1641
φ_B	0.3290	0.5006	0.5006	0.3290
φ_C	0.1641	0.2497	0.2497	0.5067

The table shows that the direct source links ($A0$ for Ann and $C0$ for Carl) have the highest marginal reliability, and in general the marginal reliability differs between agents and edges. Thus, using proportional cost sharing with respect to marginal reliability will not imply an equal split as in the case of marginal connectivity (see example 3.5 above). Cost will be shared according to the cost ratios shown in the table. For instance, if all edge costs are normalized to 1, the total loop cost of 4 will be split as $(7/6, 10/6, 7/6)$.

In section 4.2 we return to such rules in a more general network model with redundancies.

3.5 Exercises

3.1. Fishburn and Pollak (1983) focus on the proportional rule for fixed-route cost sharing. Suggest alternative rules that satisfy individual rationality. Which of the axioms characterizing the proportional rule are violated?

3.2. Concerning fixed-route cost sharing, consider the consequences of abandoning the triangular inequalities.

Also, in the above analysis it was assumed that $c_{ij} = c_{ji}$. In practice, however, this is often not the case (for example, on June 10, 2015, a Scandanavian Airlines flight from Copenhagen to Beijing costs 5,390 Danish Kroner, while the flight from Beijing to Copenhagen costs 8,541 Kroner). Examine the consequences of allowing asymmetric costs, $c_{ij} \neq c_{ji}$.

3.3. Show that every traveling salesman problem on a connected graph can be transformed into a problem on a complete graph with weights satisfying the triangular inequalities (3.5). Try to formulate the associated cost game as a mixed programming problem. (See, e.g., Tucker, 1960).

3.4. Prove that the core of the traveling salesman game is nonempty when $n \leq 3$.

3.5. (Curiel, 1997) Consider the case where for each $i \in N^0$ there exist numbers a_i, b_i such that $c_{ij} = a_i + b_j$ for all $i, j \in N^0$, with $i \neq j$. For instance, the traveling costs involved consist of a tax for entering and leaving each city.

Prove that in this case, the core of the associated traveling salesman game takes the form

$$\left\{ x \in \mathbb{R}^n \mid x_i = a_i + b_i + \lambda_i(a_0 + b_0), \, \lambda_i \in [0, 1], \sum_{N^0} \lambda_i = 1 \right\}.$$

3.6. (Estevez-Fernandez et al., 2009) Consider the situation from the fixed-route cost-allocation problem, but add to it that every time a city (node) is visited, revenue $b_i \geq 0$ is obtained. Assume that it is profit maximizing for coalition N to take the full tour, that is, we can define a profit game for a cost-minimizing tour π, where $v^\pi(N) = \sum_{i \in N} b_i - c^\pi(N)$ and for $S \subset N$, we have $v^\pi(S) = \max_{R \subseteq S} \left\{ \sum_{i \in R} b_i - c^\pi(R) \right\}$, since for some subsets it may be profit maximizing to skip some of the cities (if the cost of visiting these cities exceeds the revenue). Under which conditions is this game strategically equivalent to the fixed-route cost game? It can be shown that the core of the profit game is nonempty— provide some intuition.

3.7. Reconsider the situation in example 3.3. Try to add edges to the graph such that it becomes Eulerian. Assume that these added edges can be used by the postman for free. How does this change the game? Is the core nonempty now?

3.8. Consider the Chinese postman game (3.8) for a complete graph with six edges (four nodes) and edge costs normalized to 1. What is the total cost to be allocated? Is the complete graph weakly Eulerian? If not, argue why the core of the game (3.8) is empty.

Next, consider the graph below.

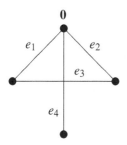

Is this graph weakly Eulerian? For edge costs normalized to 1, determine the allocation given by ϕ^E as defined in (3.9). Determine the associated four-agent game (3.8) in this case. Are all allocations in the core of this game reasonable?

3.9. (Hamers et al., 1999) Consider the Chinese postman game (3.8) and the allocation rule ϕ^E defined in (3.9). Prove that when the graph G is weakly

Eulerian, ϕ^E is a core solution. (Hint: Use that every $S \subseteq E$ can be partitioned into at most $|B(G)| + 1$ sets $S_0, \{S_b\}_{b \in B(G)}$, with $S_0 = S \cap \{E - \cup_{b \in B(G)} F_b(G)\}$ and $S_b = S \cap \{F_b(G) - \cup_{\{b' \in B(G) \cap F_b(G), b' \neq b\}} F_{b'}(G)\}$).

3.10. Consider three agents $\{1, 2, 3\}$ who all want to connect to the root 0 in the source loop (with one redundant edge) illustrated below. Suppose that there is a 10 percent chance of failure for each edge, and assume that these are stochastically independent. Moreover, let the cost of each edge be 1.

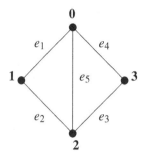

Compute the reliability for each agent and share the total cost 5 in proportion to marginal reliability. Compare with cost sharing in proportion to marginal connectivity and comment.

3.11. Continue exercise 3.10. How is the total cost of 5 allocated if we use the ex ante rule (3.15)? Once the uncertainty is resolved and we know which edges have failed, we can allocate the costs as in a deterministic situation: assume here that agents for whom a given edge can provide service (be part of a path ensuring connectivity) share its cost equally. Suggest an allocation rule based on this type of fairness idea, but in the stochastic setting where cost shares are settled ex ante.

References

Ball, M. O. (1979), "Computing Network Reliability," *Operations Research* 27: 823–838.

Ball, M. O., C. J. Colbourn, and J. S. Provan (1995), "Network Reliability," in M. O. Ball et al., (eds.), *Network Models*, vol. 7 in *Handbooks in Operations Research and Management Science.* Amsterdam: Elsevier.

Chen, H.-L., T. Roughgarden, and G. Valiant (2010), "Designing Network Protocols for Good Equilibria," *SIAM Journal on Computing* 39: 1799–1832.

Chun, Y., (1988), "The Proportional Solution for Rights Problems," *Mathematical Social Sciences* 15: 231–246.

Curiel, I. (1997), *Cooperative Game Theory and Applications: Cooperative Games Arising from Combinatorial Optimization Problems.* New York: Springer.

Derks, J., and J. Kuipers (1997), "On the Core of Routing Games," *International Journal of Game Theory* 26: 193–205.

Edmonds, J., and E. L. Johnson (1973), "Matching, Euler Tours and the Chinese Postman," *Mathematical Programming* 5: 88–124.

Engevall, S., M. Göthe-Lundgren, and P. Värbrand (1998), "The Traveling Salesman Game: An Application of Cost Allocation in a Gas and Oil Company," *Annals of Operations Research* 82: 453–471.

Estevez-Fernandez, A., P. Borm, M. Meertens, and H. Reijnierse (2009), "On the Core of Routing Games with Revenues," *International Journal of Game Theory* 38: 291–304.

Fishburn, P. C. (1990), "Fair Cost Allocation Schemes," *Social Choice and Welfare* 7: 57–69.

Fishburn, P. C., and H. O. Pollak (1983), "Fixed-Route Cost Allocation," *American Mathematical Monthly* 90: 366–378.

Golden, B., S. Raghavan, and E. Wasil (2008), *The Vehicle Routing Problem: Latest Advances and New Challenges*. New York: Springer.

Granot, D., H. Hamers, and S. Tijs (1999), "On Some Balanced, Totally Balanced and Submodular Delivery Games," *Mathematical Programming* 86: 355–366.

Hamers, H. (1997), "On the Concavity of Delivery Games," *European Journal of Operational Research* 99: 445–458.

Hamers H., P. Borm, R. van de Leensel, and S. Tijs (1999), "Cost Allocation in the Chinese Postman Problem," *European Journal of Operational Research* 118: 153–163.

Hougaard, J. L., (2009), *An Introduction to Allocation Rules*. Newe York: Springer.

Hougaard, J. L., and H. Moulin (2014), "Sharing the Cost of Redundant Items," *Games and Economic Behavior* 87: 339–352.

Hougaard, J. L. and H. Moulin (2017), "Sharing the Cost of Risky Projects," *Economic Theory*, in press New York: Springer.

Jünger, M., G. Reinelt, and G. Rinaldi (1995), "The Traveling Salesman Problem," in M. O. Ball et al., (eds.), vol. 7 in *Handbooks in Opeartions Research and Management Science*. Amsterdam: Elsevier.

Kuipers, J. (1993), "A Note on the 5-Person Traveling Salesman Game," *Zeitschrift für Operations Research* 38: 131–139.

Kwan, M. K. (1962), "Graphic Programming Using Odd or Even Points," *Chinese Mathematics* 1: 273–277.

Moulin, H., (1987), "Equal or Proportional Division of a Surplus, and Other Methods," *International Journal of Game Theory* 16: 161–186.

Potters, J., I. Curiel, and S. Tijs (1992),"Traveling Salesman Games," *Mathematical Programming* 53: 199–211.

Tamir, A. (1989), "On the Core of the Traveling Salesman Cost Allocation Game," *Operations Research Letters* 8: 31–34.

Tucker, A. W. (1960), *On Directed Graphs and Integer Programs*. Princeton, NJ: Princeton University Press.

4 General Networks

We now turn to general network structures, which can include multiple roots (sources) and sinks, grids, and other combined forms of tree and cycle structures, as illustrated by the graphs below.

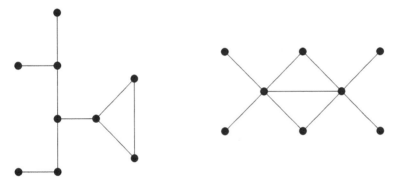

It is still fruitful to distinguish between models involving a fixed network and those involving a network design problem. In the latter case we can interpret the problem as a cooperative game and apply standard solution concepts to settle allocation issues as usual. But we now have an added complication: in general network structures, there is no guarantee of a nonempty core, and even if the core is nonempty we easily end up in situations where no core allocation is compelling. In many cases, simple allocation rules may therefore outperform the more subtle game-theoretic solution concepts.

For fixed networks the allocation problems are equally challenging. In general network structures, individual agents' network usage is particularly difficult to determine, since the network offers multiple ways for given agents to obtain their desired connectivity, possibly via redundant connections. Basically two main approaches can be used account for individual agents' liability: we can either focus directly on the agent's connectivity options and measure her potential usage of given connections (edges); or we can focus on the agent's cost responsibility for given connections, bounded by her standalone cost. Both approaches have their pros and cons, depending on the specific context.

The models we analyze encompass straightforward extensions and variations of the models from chapter 2. In fixed networks with a tree structure (but no root), agents are no longer identified by the set of nodes. Instead their individual connection demand consists of some pair of distinct nodes they want connected directly or indirectly. Thus, we can use the fact that there is a unique path between any two nodes to express agents' usage of given edges, and fairness seems to suggest that all agents using a given edge share its cost equally. In the hierarchy model it is also relatively straightforward to generalize both the family of fixed-fraction transfer rules and theorem 2.1 to the more general case of joint control (i.e., hierarchies where agents may have multiple direct superiors).

When generalizing connection demands in the MCST model, we run into two complications. First, for derived cooperative games, the situation now resembles the Steiner tree model, because some nodes may not be demanded by anybody, but are nevertheless available for obtaining the desired connectivity, so derived games may have empty cores. Second, the irreducible cost matrix is no longer well defined, so rules like the Folk solution have no obvious counterpart. To make up for this, strong separability assumptions can be imposed. The full relevance of these separability conditions will be clear when analyzing implementation issues in chapter 6. Moreover, by replacing the assumption of constant connection costs with either convex or concave connection costs, optimal networks may now include cycles in both cases (as already demonstrated in example 2.10 and at the start of chapter 3). However, adding externalities to the model in the form of capacity requirements still leads to optimal networks in the shape of spanning trees. So, in that case, the problem can basically be handled using variations of standard allocation rules, and the naturally derived cooperative game is concave. Thus, the core is relatively large, and conventional game-theoretic solutions, like the Shapley value, produce core allocations.

When agents are free to decide whether to form connections, and the societal value of resulting network structures can be assessed, we examine how such resulting network values can be fairly allocated among the agents. Since the value of connecting the same set of agents may differ depending on the specific network configuration, the model is richer than the usual TU game, but it turns out to be relatively straightforward to extend the tools of cooperative game theory to such a setup based on network value (dubbed, network games).

Finally, we will examine allocation problems arising from flows in networks: either when distributing some amount of a homogeneous resource from multiple sources to agents located at multiple sinks, given the structure of the network; or when allocating the value of a flow from a single source to a single sink between agents controlling individual arcs in the network. The operations research literature provides several algorithms to determine optimal flows, but problems dealing with fair allocation issues have been largely ignored.

4.1 Hierarchies with Joint Control

As a first example of a fixed network with a general structure, we analyze an extension of the linear hierarchy model (from chapter 2) allowing for multiple superiors as well as subordinates (see Hougaard et al., 2017). Following the approach in chapter 2, this situation can be modeled by a fixed digraph indicating the hierarchical relationship between agents, but now with potentially multiple sources (bosses exercising joint control over certain subordinates) as well as multiple terminal nodes (lowest-ranked agents).

4.1.1 The Model

A *joint control revenue-sharing problem* is defined as a triple (N, r, D), where N is a nonempty finite set of agents, r is a profile specifying the revenue r_i of each agent $i \in N$, and D is a dominance structure mapping each agent $i \in N$ to her immediate superiors $D(i) \subset N$ (the predecessors of i in the digraph) with the convention that $D(i) = \emptyset$ if i is a boss. We assume that the digraph induced by D is connected and (undirected) cycle free (i.e., the associated undirected graph contains no cycles). Let $\mathcal{J}$ denote the set of such problems.

Since the hierarchy is cycle free, deleting any edge ij leads to two components dubbed the i- and the j-components and denoted by G_{ij}^i and G_{ij}^j, respectively.

Given a problem (N, r, D), an allocation is a vector $x \in \mathbb{R}^N$ satisfying budget balance. An *allocation rule* ζ is a mapping that, for each problem (N, r, D), assigns an allocation $x = \zeta(N, r, D)$. As in the linear hierarchy model, the allocation rule is assumed to be anonymous, that is, for each bijective function $g : N \to N'$, $\zeta_{g(i)}(N', r', S') = \zeta_i(N, r, S)$, where $r'_{g(i)} = r_i$ and $S'(g(i)) = g(S(i)) = \{g(s) : s \in S(i)\}$ for each i.

The family of fixed-fraction transfer rules (defined in (2.3) for linear hierarchies, and in (2.18) and (2.19) for branch hierarchies) generalizes easily to the joint-control setting by transferring an equal split of the accumulated surplus of a given agent i to all immediate superiors $j \in D(i)$. That is, a lowest-ranked agent (i.e., i for which $D^{-1}(i) = \emptyset$) gets payoff λr_i, while a predecessor $j \in D(i)$ gets payoff

$$\lambda \left(r_j + \sum_{k \in D^{-1}(j)} \frac{(1 - \lambda)r_k}{|D(k)|} \right), \tag{4.1}$$

and so forth. Denote this generalized family of fixed-fraction transfer rules by $\{\zeta^\lambda\}_{\lambda \in [0,1]}$. Clearly, every member of the family $\{\zeta^\lambda\}_{\lambda \in [0,1]}$ is anonymous.

4.1.2 Axioms and Characterization

Most axioms from the linear hierarchy model (see section 2.1.1) have a natural extension (dubbed "j-") to the joint control model.

j-Highest rank revenue independence   For each $(N, r, D) \in \mathcal{J}$, each $i \in N$ such that $D(i) = \emptyset$, and each $\hat{r}_i \in \mathbb{R}_+$,

$$\zeta_{N \setminus \{i\}}(N, r, D) = \zeta_{N \setminus \{i\}} \left(N, (r_{-i}, \hat{r}_i), D \right).$$

That is, the size of a given boss's revenue has no influence on the payoff of any other agent.

j-Highest rank splitting neutrality   For each $(N, r, D) \in \mathcal{J}$ and each $i \in N$ such that $D(i) = \emptyset$, let (N', r', D') be such that $N' = N \cup \{k\}$, $D'(i) = k$, $D = D'$, otherwise, $r_i = r'_k + r'_i$, and $r'_{N \setminus \{i,k\}} = r_{N \setminus \{i\}}$. Then

$$\zeta_{N \setminus \{i,k\}}(N', r', D') = \zeta_{N \setminus \{i\}}(N, r, D).$$

That is, if a boss splits her revenue among multiple aliases, no other agent's payoff is affected.

j-Scale covariance   For each $(N, r, D) \in \mathcal{J}$, and each $\alpha > 0$,

$$\zeta(N, \alpha r, D) = \alpha \zeta(N, r, D).$$

That is, if revenues are scaled up or down, so are the payoffs.

However, the lowest rank consistency axiom is no longer directly meaningful and must be strengthened. Consider an agent i and one of his immediate subordinates j. Deleting the j component and transferring any surplus from that component to i should leave the payoffs of all agents in the i component unchanged. Clearly, this implies lowest rank consistency, since the j component may consist of agent j alone.

Component consistency   Let $(N, r, D) \in \mathcal{J}$ be a problem where $i \in D(j)$ with resulting payoff $x = \zeta(N, r, D)$. Define a new problem (N', r', D') by deleting the j component G_{ij}^j and with $r_i' = r_i + \sum_{k \in G_{ij}^j} (r_k - x_k)$, $r_h' = r_h$ otherwise, for the remaining agents. Then

$$\zeta_h(N', r', D') = \zeta_h(N, r, D)$$

for all $h \in G_{ij}^i$.

The following axiom is new compared to the scenario with linear and branch hierarchies: in a situation where a given agent has several bosses who are not bosses for any other agent, merging these bosses into one agent will not change the payoff of the remaining (unmerged) agents.

Top merger Consider a problem $(N, r, D) \in \mathcal{J}$ and a given (non-boss) agent j with $|D(j)| \geq 2$ and $D(k) = \emptyset$ for all $k \in D(j)$. Define a new j-problem (N', r', D'), where $N' = (N \setminus D(j)) \cup k'$; $r_{k'}' = \sum_{k \in D(j)} r_k$, and $r' = r$ otherwise; $D'(k') = \emptyset$, $D'(j) = k'$, and $D' = D$ otherwise. Then

$$\zeta_l(N, r, D) = \zeta_l(N', r', D')$$

for all $l \in N \setminus D(j)$.

With these extended axioms in place, Hougaard et al. (2017) extend theorem 2.1 to (cycle-free) joint-control revenue-sharing problems.

Theorem 4.1 An allocation rule ζ satisfies j-highest rank revenue independence, j-highest rank splitting neutrality, j-scale covariance, component consistency, and top merger if and only if $\zeta \in \{\zeta^\lambda\}_{\lambda \in [0,1]}$.

Proof By theorem 2.1 there exists a λ such that $\zeta = \zeta^\lambda$ for up to two-agent problems. Suppose there is λ such that $\zeta = \zeta^\lambda$ for all problems with up to k agents, $k \geq 2$, and consider the subfamily of problems with $k + 1$ agents. Take an arbitrary agent i for which $D^{-1}(i) = \emptyset$. We claim that $\zeta_i = \lambda r_i$: indeed, by repeated use of component consistency, we can construct a new problem for which all agents other than i have a unique linear path to i such that i's payoff is unchanged. By repeated use of top merger and j-highest rank splitting neutrality, we obtain a new (two-agent) problem for which agent i gets $\zeta_i = \lambda r_i$.

Now consider agents $j \in D(i)$ and their j-component G_{ij}^j. We claim that $\sum_{h \in G_{ij}^j} x_h = \sum_{h \in G_{ij}^j} r_h + \frac{(1-\lambda) r_i}{|D(i)|}$. Indeed, by repeated use of component consistency, top merger, and j-highest rank splitting neutrality, we can reduce the j-components to a single agent, where this agent receives the same payoff as the entire j-component did before. By j-highest rank revenue independence and anonymity, we thus verify the claim.

From the claim verified above, we have $\sum_{h \in G^i_{ij}} x_h = \sum_{h \in G^i_{ij}} \zeta^\lambda_h(N, r, D)$. For a given $j \in D(i)$, by component consistency we can add the surplus of the i-component ($\sum_{h \in G^i_{ij}} (r_h - x_h)$) to j and then eliminate the i-component. By the induction hypothesis, the payoff of an arbitrary agent $h \in G^j_{ij}$ is $x_h = \zeta^\lambda_h(N, r, D)$. ∎

Example 4.1 (Illustrating the model and computing ζ)   Consider a situation where two bosses, 5 and 6, have agent 4 as a common direct subordinate; agent 4 is the superior of agent 3, who has two subordinates, agents 1 and 2; agent 2 also has agent 7 as boss (recall that a boss is an agent without superiors, i.e., a source in the digraph).

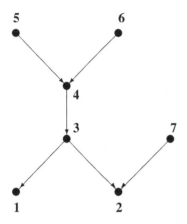

Thus, the (cycle-free) digraph above is induced by the dominance structure: $D(5) = D(6) = D(7) = \emptyset$; $D(4) = \{5, 6\}$; $D(3) = \{4\}$; $D(2) = \{3, 7\}$; $D(1) = \{3\}$.

Let revenues be normalized such that $r_i = 1$ for all i. Using a fixed-fraction transfer rule with $\lambda = 0.5$, we get payoffs

$$\zeta^{0.5} = \left(\frac{1}{2}, \frac{1}{2}, \frac{7}{8}, \frac{15}{16}, \frac{47}{32}, \frac{47}{32}, \frac{5}{4} \right).$$

In this case agents 1 and 2 keep half of their revenue, agent 1 transfers his surplus of 0.5 to agent 3, while the surplus of agent 2 is split equally between superiors 3 and 7 (this leaves the boss, agent 7, with payoff $(1 + 0.25 =)$ 1.25). Agent 3 keeps half of her own revenue plus the total surplus transferred from agents 1 and 2 (which is 0.75), yielding 7/8. Agent 4 keeps half of his own revenue plus the surplus of 7/8 transferred from agent 3, yielding 15/16. Bosses

5 and 6 split the surplus of $15/16$ transferred from agent 4 equally, yielding a payoff of $47/32$ to both.

Note that merging bosses 5 and 6 into one agent with a total revenue of 2 does not change the payoff of any of the other agents (as per the axiom top merger).

Even though theorem 4.1 characterizes the family of transfer rules on the domain of (undirected) cycle-free graphs, the fixed-fraction transfer rules ζ^λ are well defined for any hierarchical digraph without *directed* cycles. That is, hierarchical relationships leading to undirected cycles (e.g., between agents $\{i, j, h\}$, where $D(i) = \emptyset$, $D(j) = \{i\}$, $D(h) = \{i, j\}$) do not prevent the use of fixed-fraction transfer rules.[1] The fixed-fraction transfer rules thus can be seen as regulating a flow in a general digraph (without directed cycles). In this case, the flow is of revenue from multiple sources to multiple sinks when agents (here represented by nodes) are allowed to withhold a fraction of the flow passing through their node.

Finally, if the joint-control problem is construed as a game with permission structure, where an underlying additive game in revenues is restricted by the permission structure of the hierarchy, there is now a difference between the conjunctive and the disjunctive approach (see section 2.1.1). Consider the graph in example 4.1. Any coalition containing, say, agent 2, must also contain agents 3, 4, 5, 6, and, 7, in order to be "cooperative" according to the conjunctive approach. However, according to the disjunctive approach, coalitions like $\{2, 7\}$, $\{2, 3, 4, 5\}$, and $\{2, 3, 4, 6\}$ will also be able to cooperate. Thus, there is now a difference between applying solution concepts, such as the Shapley value, to these two forms of restricted games.

4.2 Networks with Redundant Connections

Following Hougaard and Moulin (2014), let us now consider a situation where networks are fixed and undirected. Moreover, the structure of the network has no restrictions, so it may contain redundant links that no agent needs to obtain connectivity. Hence, the network may not be efficient in a cost-minimizing sense. Edges are assumed to be public goods (i.e., they can be used by all agents without rivalry). Agents have connection demands in the form of nodes that must be connected (either directly or indirectly via other nodes). Demands are binary in the sense that an agent is satisfied if the desired connectivity is

---

1. Only cyclic relationships like, $D(i) = \{h\}$, $D(j) = \{i\}$, $D(h) = \{j\}$), cannot be handled by the transfer rules.

obtained and dissatisfied if not. Each agent's connectivity constraint can therefore be represented by a set of subgraphs that satisfy the agent's connection demand.

Every edge has a cost, and the total cost of the network has to be covered by the network users, including the cost of redundant edges. Agents prefer a low-cost share, but other differences in their individual preferences are irrelevant, making demand fully inelastic. As such, the situation can be viewed as one where a group of agents takes over an existing network structure en bloc, which is not necessarily designed to meet their demands, but nevertheless provides everybody with their desired connectivity. As usual, we search for fair ways to share the total network cost among users.

A straightforward and simple approach ignores most of the information contained by the connectivity constraints and individual edge costs. It allocates the total network cost in proportion to the standalone cost of each agent (here being the minimal cost needed to satisfy the agent's connectivity demand). This follows the spirit of Fishburn and Pollak's suggestion for the fixed-route cost-allocation problem (see section 3.1), where the network is given by a cycle, but additional information about round-trip costs for each city is available. In fact, the current model allows us to determine the standalone cost for any coalition of agents and thus to define an associated cooperative game, noting that the cost of the grand coalition is the total cost of the network (which will not be cost minimizing in the case of redundancy). Allocation of total cost can therefore be obtained using solution concepts for cooperative games as well.

None of these solutions will be additive in edge costs, though. In addition, since the network is fixed and has to be accessed en bloc, all coalitional values (except for the grand coalition) are counterfactual. Moreover, representing the problem as a cooperative game ignores much of the rich information conveyed by agents' connectivity constraints (i.e., information revealing how the agents actually benefit from the presence of the various edges in the network). So to base a normative analysis on hypothetical standalone costs or a derived cooperative game is not a natural starting point. We therefore take an alternative approach related to the idea of cost protocols for fixed networks, which was introduced in section 3.4.

4.2.1 The Model

Let $\mathcal{M}$ and $\mathcal{G}$ be infinite sets of potential agents and connected graphs (networks), respectively. A given problem concerns a finite set of agents $M \subset \mathcal{M}$, and a particular connected graph $G(N, E) \in \mathcal{G}$. Agents $i \in M = \{1, \dots, m\}$

have connection demands $(a_i, b_i) \in N \times N$ in the form of pairs of (distinct) nodes in G that they want to connect. Edges are undirected public goods.

A cost vector $c \in \mathbb{R}_+^E$ specifies the cost c_e of every edge $e \in E$ (for instance, edge-specific maintenance costs). The agents in M must share the total cost of the entire network $c_E = \sum_{e \in E} c_e$, even if only a subset of the edges are enough to satisfy all connection demands, and consequently some edges are redundant.[2]

Connection demands can be satisfied by different paths in the network.[3] We can then model the connectivity needs of agent $i \in M$ by a collection of subgraphs $\mathcal{D}^i \subseteq 2^E \setminus \varnothing$, where each $D \in \mathcal{D}^i$ is a subgraph in G satisfying i's connection demand, i.e., the subgraph D contains at least one path connecting nodes $(a_i, b_i) \in N \times N$, directly or indirectly. We call $\mathcal{D}^i$ the *service constraint* of agent i, and i is *served* if $\mathcal{D}^i \neq \varnothing$. Denote by $\overline{\mathcal{D}}^i \subseteq \mathcal{D}^i$ the set of all minimal subgraphs serving i, that is, subgraphs for which deleting any one edge will lead to service failure (corresponding to the set of all paths satisfying the connection demand of agent i).

Example 4.2 (Illustrating service constraints) Consider the graph $G = (N, E)$ below with four nodes $N = \{1, 2, 3, 4\}$ and four edges $E = \{a, b, c, d\}$.

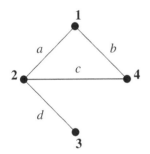

Assume there are two agents, Ann and Bob. Ann wants to connect nodes 1 and 4, while Bob wants to connect nodes 1 and 3. Thus, the minimal subgraphs of Ann are $\overline{\mathcal{D}}^{Ann} = \{b, ac\}$, while Bob's are $\overline{\mathcal{D}}^{Bob} = \{ad, bcd\}$, so all subgraphs containing elements from $\overline{\mathcal{D}}^{Ann}$ and $\overline{\mathcal{D}}^{Bob}$ constitute their respective service constraints:

$$\mathcal{D}^{Ann} = \{b, ab, bc, bd, ac, abc, acd, abd, bcd, abcd\}$$

---

2. Note that it is not always possible to tell exactly which edges are redundant in a given graph, for example, if all agents in a cyclic graph demands 1-connectivity to a source.

3. Recall that paths are sequences of edges in which no node is visited twice except perhaps the first node (which may also be the last in a cyclic path).

and

$$\mathcal{D}^{Bob} = \{ad, abd, acd, bcd, abcd\}.$$

The graph clearly contains a redundant edge, since we can satisfy both Ann's and Bob's demand deleting either a, b or c (by choosing b and bcd; ac and ad; and b and ad, respectively).

A *cost-allocation problem* is a list $(M, G, \{\mathcal{D}^i\}_{i \in M}, c)$. Denote by $\Gamma^{M,G}$ the set of cost-allocation problems with agent set M and graph G. Let Γ be the set of all allocation problems: $\Gamma = \bigcup_{M \subset \mathcal{M}, G \subset \mathcal{G}} \Gamma^{M,G}$.

A *cost-allocation rule* ϕ is a mapping with domain Γ such that

$$\phi(M, G, \{\mathcal{D}^i\}_{i \in M}, c) \in \mathbb{R}_+^M$$

for any problem in $\Gamma^{M,G}$, satisfying budget balance:

$$\sum_{i \in M} \phi_i(M, G, \{\mathcal{D}^i\}_{i \in M}, c) = c_E.$$

From the outset we assume that relevant cost-allocation rules ϕ must be *additive in edge costs*.

Cost additivity   For any two problems that differ only in their edge costs,

$$\phi(M, G, \{\mathcal{D}^i\}_{i \in M}, c + c') = \phi(M, G, \{\mathcal{D}^i\}_{i \in M}, c) + \phi(N, R, \{\mathcal{D}^i\}_{i \in M}, c'). \tag{4.2}$$

If a rule is cost additive, we have

$$\phi(M, G, \{\mathcal{D}^i\}_{i \in M}, c) = \sum_{e \in E} c_e \times \varphi(M, G, \{\mathcal{D}^i\}_{i \in M}, e), \tag{4.3}$$

where the vector $\varphi(M, G, \{\mathcal{D}^i\}_{i \in M}, e) \in \Delta(M)$ (the M-simplex) assigns relative shares to each agent in the division of the edge cost c_e. The coordinate $\varphi_j(M, G, \{\mathcal{D}^i\}_{i \in M}, e)$ is called agent j's *cost ratio* of edge e (given the profile $\{\mathcal{D}^i\}_{i \in M}$ of service constraints). Cost-additive rules are also known as *protocols* in the computer science literature.

Focusing on cost additive rules implies that a fair cost share of a given edge e does not depend on its cost or the cost of other edges, but rather on the structure of the service constraints of all agents (i.e., the way agents are able to obtain connectivity using the various edges).

Consider the following example of a cost-additive rule defined for cases with no redundant edges in the network.

Example 4.3 (The equal-need rule in the nonredundant case)   A problem
is called nonredundant if removing any edge will lead to connectivity failure
for at least one agent. Let $N(e) = \{i \in M \mid E \setminus \{e\} \notin \mathcal{D}^i\}$ be the set of agents
who need the edge e in order to get connectivity. In a nonredundant problem,
every edge is critical for at least one agent, so $N(e) \neq \emptyset$ for all $e \in E$. Now,
as suggested in Moulin and Laigret (2011), a natural cost-additive rule shares
the cost of each edge equally among agents, for whom the edge is critical, and
payments are found by adding up over edges,

$$\phi_i^{EN} = \sum_{e:i \in N(e)} \frac{c_e}{|N(e)|}, \tag{4.4}$$

for all $i \in M$. This rule is dubbed the equal-need rule. Notice that agents do not
pay for edges that are not critical to them. Hence, if no edges are critical to a
given agent, this agent free rides. For instance, consider two agents on a source
loop: Ann demands 2-connectivity to the source, while Bob is satisfied by
1-connectivity; according to the equal-need rule, Bob gets connected for free.
Obviously, it does not seem reasonable that agents are allowed to free ride if
no edge is critical to them. After all, they are part of the network and obtain
service from it.

It is shown in Moulin and Laigret (2011) that the equal-need rule is the
only relevant cost-additive rule that satisfies the (standalone) core conditions
of the associated cooperative cost game (where the cost of any coalition $S \subseteq M$
is simply given by the smallest cost of satisfying the connection demand of
all members in S). In light of the free-rider problem mentioned above, there
are good reasons to go beyond the core conditions if we insist on using cost-
additive allocation rules, even in case of nonredundancy.

Consider the graph $G = (N, E)$ below with four nodes $N = \{1, 2, 3, 4\}$,
three edges $E = \{b, c, d\}$, and two agents Ann and Bob. Ann wants to connect nodes 1 and 4, while Bob wants to connect nodes 1 and 3. In this sense,
node 1 is a common root.

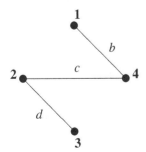

The derived cost-allocation problem is clearly nonredundant, since deleting edge b will lead to connectivity failure for both Ann and Bob, while deleting either of the edges c or d will lead to connectivity failure for Bob. Applying the equal-need rule makes Ann and Bob split the cost of edge b equally, while the cost of edges c and d are covered by Bob alone. So the equal-need rule coincides with the serial rule on the domain of nonredundant rooted fixed-tree allocation problems (see section 2.1.2.).

4.2.2   Cost Ratios and Liability Indices

Cost ratios capture the relative importance of a given edge for a given agent, relative to the full profile of service constraints for all agents. This suggests some natural restrictions on φ. The liability of agent i for edge e should be largest when e is part of every path satisfying i's connection demand, that is, when e is *critical* to i. Let K^i be the set of edges critical to agent i. Moreover, we say that the edge e is *relevant* for agent $i \in M$ if it is part of at least one path in $\overline{\mathcal{D}}^i$, in which case it is fair to hold i liable for some of e's cost. On the contrary, if edge e is *irrelevant* to agent i, then e has no impact on whether i gets connected or not, so i should not pay for e. Let H^i be the set of edges that are relevant for agent i. Finally, it seems reasonable to require that the agents are equally liable for edges that are irrelevant to everybody.

Formally, a cost ratio index $\varphi(M, G, \{\mathcal{D}^i\}_{i \in M}; e) \in \Delta(M)$ must satisfy the following four assumptions:

1. If edge e is critical to agent i at $\widetilde{\mathcal{D}}^i$, then i's cost ratio is largest:

$$\varphi(\widetilde{\mathcal{D}}^i, \{\mathcal{D}^j\}_{j \in M \setminus i}; e) \geq \varphi(\mathcal{D}^i, \{\mathcal{D}^j\}_{j \in M \setminus i}; e) \text{ for all } \mathcal{D}^i \subseteq 2^E \setminus \varnothing$$

(for simplicity, M and G are omitted in the argument of φ).

2. If edge e is relevant to agent i at $\mathcal{D}^i$ then i pays something of its cost:

$$e \in H^i \Rightarrow \varphi_i(\{\mathcal{D}^j\}_{j \in M}; e) > 0.$$

3. If edge e is irrelevant for agent i at $\mathcal{D}^i$ but is relevant to some other agent(s), then i is not charged for e:

$$e \in \{E \setminus H^i\} \cap \{\cup_{j \in M \setminus \{i\}} H^j\} \Rightarrow \varphi_i(\{\mathcal{D}^j\}_{j \in M}; e) = 0.$$

4. If edge e is irrelevant for all agents, its cost is divided equally:

$$e \in E \setminus \cup_{j \in M} H^j \Rightarrow \varphi_i(\{\mathcal{D}^j\}_{j \in M}; e) = \frac{1}{m} \text{ for all } i.$$

The main idea is now to focus on cost ratios that are defined proportional to the agent's liability for a given edge e, that is, takes the form

$$\varphi_i(\{\mathcal{D}^j\}_{j\in M}, e) = \frac{\theta(\mathcal{D}^i, e)}{\sum_{j\in M} \theta(\mathcal{D}^j, e)} \text{ for all } i, \tag{4.5}$$

where $\theta(\mathcal{D}^i, e)$ is agent i's *liability index* for edge e at $\mathcal{D}^i$. This defines a cost ratio index if

$$\theta(\mathcal{D}^i, e) \geq 0; \ \theta(\mathcal{D}^i, e) \text{ is maximal w.r.t. } \mathcal{D}^i \text{ if } e \in K^i, \tag{4.6}$$

$$\theta(\mathcal{D}^i, e) > 0 \text{ if } e \in H^i; \ \theta(\mathcal{D}^i, e) = 0 \text{ if } e \notin H^i, \tag{4.7}$$

with the convention $\frac{0}{0} = \frac{1}{m}$.

Example 4.4 (Two liability indices) The simplest example of a liability index meeting (4.6) and (4.7) only checks whether an edge is relevant:

$$\theta^0(\mathcal{D}^i, e) = 1 \text{ if } e \in H^i, \ \theta^0(\mathcal{D}^i, e) = 0 \text{ otherwise.} \tag{4.8}$$

This egalitarian (liability) index is obviously very crude and basically ignores the rich information of the service constraints. In effect, a cost-additive allocation rule based on cost ratios with this egalitarian form of liability index implies that the cost of a given edge c_e is split equally among all the agents in M for whom edge e is part of some path that satisfies their connection demand.

But liability indices may also be quite sophisticated by construction, as for the following probabilistic liability index $\theta^P(\mathcal{D}^i, e)$: Consider a given agent i with service constraint $\mathcal{D}^i$ and an ordering σ of edges E. Interpret σ as the order in which edges are paid, and stop paying once a minimal serving set of agent i is completely paid for. Now, loosely speaking, $\theta^P(\mathcal{D}^i, e)$ is the probability that we have to pay for edge e, when σ is drawn with uniform probability. However since several minimal serving sets may be completely paid at the same edge, we need to select one of them as follows. An ordering σ of E induces a lexicographic ordering $\succ_\sigma$ of $2^E \setminus \emptyset : D \succ_\sigma D'$ holds if the first edge in D (w.r.t. σ) precedes the first edge in D'; or if these two coincide and the second edge in D precedes that in D', and so on. Given the service constraints $\mathcal{D}^i$, the (minimal) serving sets completed first in the ordering σ are defined as $\mathcal{F}^i(\sigma) = \arg\min_{D^i \in \mathcal{D}^i}\{\max_{e\in D^i} \sigma(e)\}$. In the set $\mathcal{F}^i(\sigma)$, we select $D^i(\sigma)$ such that $D^i(\sigma) \succ_\sigma D^i$ for all $D^i \in \mathcal{F}^i(\sigma)$. Finally,

$$\theta^P(\mathcal{D}^i, e) = \frac{|\{\sigma \mid e \in D^i(\sigma)\}|}{|E|!}, \tag{4.9}$$

that is, the probability that a given edge e is in $D^i(\sigma)$. Clearly (4.6) and (4.7) hold true. The probabilistic index is by design heavier on small minimal subgraphs than on large ones, a somewhat less compelling feature that is not warranted by the primitives of the model.

Recall the situation in example 4.2 above. Clearly, all edges are relevant for Bob, so $\theta^0(\mathcal{D}^{Bob}, e) = 1$ for $e \in \{a, b, c, d\}$, while only edges a, b, c are relevant for Ann, so $\theta^0(\mathcal{D}^{Ann}, e) = 1$ for $e \in \{a, b, c\}$, and $\theta^0(\mathcal{D}^{Ann}, d) = 0$. Thus, using proportional cost ratios would lead to an equal split of costs c_a, c_b, and c_c, while Bob alone pays c_d for edge d.

Computing the probabilistic index θ^P is a bit more complicated. Consider Ann. Since edge d is irrelevant, $\theta^P(\mathcal{D}^{Ann}, d) = 0$. Since edge b becomes the first minimal subgraph to serve Ann in 16 out of $4! = 24$ possible orders σ of the four edges, $\theta^P(\mathcal{D}^{Ann}, b) = \frac{16}{24} = \frac{2}{3}$. Moreover, since edges a and c become the first minimal subgraph to serve Ann in the remaining 8 out of 24 cases, we get $\theta^P(\mathcal{D}^{Ann}, a) = \theta^P(\mathcal{D}^{Ann}, c) = \frac{8}{24} = \frac{1}{3}$.

Now consider Bob. Since edge d is critical to Bob, $\theta^P(\mathcal{D}^{Bob}, d) = 1$. For the other edges, we need to make use of the lexicographic order. For instance, for $\sigma = abcd$, both minimal subgraphs ad and bcd end with edge d, but since ad includes the first element, the subgraph ad is chosen; that is, $ad \succ_\sigma bcd$. As such, edge a appears in 14 out of 24 cases where the minimal subgraph ad is chosen, while edges b and c appear in 10 out of 24 cases where the minimal subgraph bcd is chosen. Thus, $\theta^P(\mathcal{D}^{Bob}, a) = \frac{14}{24} = \frac{7}{12}$ and $\theta^P(\mathcal{D}^{Bob}, b) = \theta^P(\mathcal{D}^{Bob}, c) = \frac{10}{24} = \frac{5}{12}$. Again, using proportional cost ratios yield cost shares for Ann,

$$\varphi^{Ann} = \left(\frac{4}{11}, \frac{8}{13}, \frac{4}{11}, 0 \right),$$

and consequently $1 - \varphi^{Ann}$ for Bob.

The example above presents two types of liability indices at opposite ends of a complexity scale. A third index presents itself rather naturally. The more minimal subgraphs contained by an agent's service constraint, the more *flexible* the agent will be in her connection demand, simply because she can obtain the desired connection in more ways. When agents have more ways to obtain connectivity, it seems reasonable that they should be less liable for each individual edge in the minimal subgraphs. On the other hand, the more of these minimal subgraphs that contain a given edge, the more central that edge will be for the agent, and therefore the more liable the agent should be for this edge. These two observations can be combined using the so-called *counting liability index*, defined for each agent $i \in M$ and edge $e \in E$ as

$$\theta^1(\mathcal{D}^i, e) = \frac{|\overline{\mathcal{D}}^i(e)|}{|\overline{\mathcal{D}}^i|}, \tag{4.10}$$

where $\overline{\mathcal{D}}^i(e) = \{D \in \overline{\mathcal{D}} \mid e \in D\}$ is the set of agent i's minimal subgraphs containing the edge e. In other words, the counting index is the ratio between the number of minimal subgraphs containing the edge e and the total number of minimal subgraphs for agent i. The counting index clearly satisfies properties (4.6) and (4.7). It is, in fact, equivalent to the relative marginal contribution index (defined in (3.11)) for the nodes that agent i wants connected, provided by edge $e \in E$.

Example 4.4 (continued: The counting liability index) Recall the situation in example 4.2 above. Using the counting liability index (4.10), Ann has index values

$$\theta^1(\mathcal{D}^{Ann}, e) = \frac{1}{2} \text{ for } e \in \{a, b, c\},$$

and $\theta^1(\mathcal{D}^{Ann}, d) = 0$. Similarly, Bob has index values

$$\theta^1(\mathcal{D}^{Bob}, e) = \frac{1}{2} \text{ for } e \in \{a, b, c\},$$

and $\theta^1(\mathcal{D}^{Bob}, d) = 1$. Thus, using proportional cost ratios, Ann and Bob split the cost of edges a, b, c equally, while Bob alone pays for edge d.

Note that Bob wants to connect node 3 to node 1. If we look at the relative marginal connectivity (see (3.11)) of node 3 (using node 1 as root) provided by the edge d, we have

$$\theta_3^{rc}(G, d) = \frac{c_3(G) - c_3(G - d)}{c_3(G)} = \frac{2 - 0}{2} = 1,$$

which is identical to Bob's counting index for edge d. Similarly for edges $e \in \{a, b, c\}$ we have

$$\theta_3^{rc}(G, e) = \frac{c_3(G) - c_3(G - e)}{c_3(G)} = \frac{2 - 1}{2} = 0.5,$$

also identical to Bob's counting index of the respective edges.

Example 4.5 (The airport problem and other fixed-tree problems) Applying the counting liability index to the nonredundant case of the airport problem (from section 2.1.1.), all agents have index value 1 if the edge is relevant (here all relevant edges are critical) and 0 if not. Therefore, allocating costs in proportion to the counting index coincides with using the equal need and serial rules in this case, which again coincide with the Shapley value of the associated airport game.

For fixed trees in general, where agents demand connection between any two nodes, every agent i is characterized by the unique path p_i connecting these two nodes. Thus, $\theta_i^1(e) = 1$ if $e \in p_i$, and 0 otherwise. Consequently, if agents share costs in proportion to the counting liability index, the cost of edge e is shared equally among agents using e to get connection (and c_e is split equally among all agents if e is redundant).

A cost ratio in the proportional form (4.5) based on the counting liability index (4.10) turns out to be a prominent member of a larger family of cost ratios that can be characterized by a set of compelling fairness properties.

4.2.3 Axioms for Cost-Ratio Indices

Two standard requirements are anonymity (i.e., a cost-ratio index should be oblivious to the name of agents) and neutrality (i.e., a cost ratio index should be oblivious to the labeling of edges).

Anonymity   If σ is a permutation of agents in M, then

$$\varphi_j(\{\mathcal{D}^i\}_{i \in M}, e) = \varphi_{\sigma(j)}(\{\mathcal{D}^{\sigma(i)}\}_{i \in M}, e) \text{ for all } j \in M.$$

Neutrality   If τ is a permutation of edges in E, then

$$\varphi(\{\mathcal{D}^i\}_{i \in M}, e) = \varphi(\{\tau(\mathcal{D}^i)\}_{i \in M}, \tau(e)).$$

The next two properties are also well known in the axiomatic literature.

Consistency   For any $M \in \mathcal{M}$, $G \in \mathcal{G}$, $i \in M$, and $e \in E$, we have

$$\varphi_{-i}(M, G, \{\mathcal{D}^j\}_{j \in M}, e) = [1 - \varphi_i(M, G, \{\mathcal{D}^j\}_{j \in M}, e)]$$

$$\times \varphi(M \setminus \{i\}, G, \{\mathcal{D}^j\}_{j \in M \setminus i}, e)$$

(both sides in $\mathbb{R}_+^{M \setminus \{i\}}$).

In words, removing an agent and reducing costs accordingly, the relative cost shares of the remaining agents should be the same in the reduced and in the original problem. Any cost-ratio index taking the simple proportional form (4.5) for some liability index $\theta(\mathcal{D}^i, a)$ is clearly anonymous and consistent; it is also neutral if θ is.

Now, consider K *isomorphic* problems $(M, G_k, \{\mathcal{D}_k^j\}_{j \in M})$: that is, the graphs G_k are identical and disjoint, and there is a bijection $\tau_{kk'}$ from G_k into $G_{k'}$ sending $\mathcal{D}_k^i$ into $\mathcal{D}_{k'}^{'i}$, and such that $\tau_{k'k''} \circ \tau_{kk'} = \tau_{kk''}$ for all k, k', and k''. In other words, we take a problem and clone it into K exact replica. For instance, think of the airport problem from chapter 2: we replicate the runway and need to share the cost of these several identical runways. It seems natural to require that the way we share the cost of a runway is independent of the existence of duplicate runways.

Replication For $(M, \widetilde{G}, \{\widetilde{\mathcal{D}}^j\}_{j \in M})$, where $\widetilde{G} = \cup_k G_k$, and $\widetilde{\mathcal{D}}^j = \cup_k \mathcal{D}_k^j$, we have $\varphi(G_1, \{\mathcal{D}^j\}_{j \in M}, e) = \varphi(\widetilde{G}, \{\widetilde{\mathcal{D}}^j\}_{j \in M}, e)$ for all $e \in G_1$.

Cost-ratio indices based on the three liability indices discussed above (i.e., the egalitarian, counting, and probabilistic indices) all satisfy replication.

The final axiom is related to the connectivity constraints. Consider two problems, $P = (M, G, \{\mathcal{D}^j\}_{j \in M}) \in \Gamma^{M,G}$ and $P' = (M, G', \{\mathcal{D}'^j\}_{j \in M}) \in \Gamma^{M,G'}$, such that G' contains one more edge e' than does G. We call edge e' in P' *supplementary to agent i's connectivity needs in P*, if e' needs to be added to *some* of i's service constraints in P, but is no substitute for any initial constraint:

$$\text{for all } D \subseteq G' : D \in \mathcal{D}'^i \Rightarrow D \setminus \{e'\} \in \mathcal{D}^i, \tag{4.11}$$

$$\text{for all } D \subseteq G : D \in \mathcal{D}^i \Rightarrow D \cup \{e'\} \in \mathcal{D}'^i. \tag{4.12}$$

This implies that if D is a minimal subgraph in P, then either D or $D \cup \{e'\}$ is a minimal subgraph in P' as well, so a one-to-one correspondence exists between $\overline{\mathcal{D}}^i$ and $\overline{\mathcal{D}}'^i$, with some of the new sets including e'. Thus, adding e' does not create new connectivity options for agent i.

Irrelevance of supplementary edges If edge e' in P' is supplementary to agent i's connectivity needs in P, for all $i \in M$, then $\varphi(G, \{\mathcal{D}^i\}_{i \in M}, e) = \varphi(G', \{\mathcal{D}'^i\}_{i \in M}, e)$ for all $e \in E$.

In other words, adding a supplementary edge does not affect the cost ratio of any original edge. These ratios may change only when e' creates new service opportunities. Since for all agents, the number of minimal subgraphs is unchanged by adding a supplementary edge, it is clear that cost ratios based on the counting liability index satisfy irrelevance of supplementary edges.

Example 4.6 (More on irrelevance of supplementary edges) It seems obvious that if we add an edge that nobody can use, it should not affect the way costs are shared for the original edges, but irrelevance of supplementary edges also has more interesting implications.

We say that edges a, b, are complements if no agent needs one edge and not the other:

for all $i \in M$ and all $D^i \in \overline{\mathcal{D}}^i : \{a, b\} \subseteq D^i$ or $\{a, b\} \cap D^i = \emptyset$.

This could, for instance, be the case if all agents need to connect nodes in two different components, and edges a and b form a bridge (bottleneck) between these two components. Irrelevance of supplementary edges here implies that whether we consider the complement edges (a and b) as two separate edges with their separate cost, or as one "merged" edge with a cost equal to the sum of the individual edge costs, does not change the profile of individual cost ratios.

4.2.4 Characterization Results

The following family of liability indices generalizes the indices θ^0 and θ^1 in (4.8) and (4.10). For any number $\pi \geq 0$, any edge e, and connectivity constraints $\mathcal{D}^i$, let

$$\theta^\pi (\mathcal{D}^i, e) = \left(\frac{|\overline{\mathcal{D}}^i(e)|}{|\overline{\mathcal{D}}^i|} \right)^\pi$$

(with the convention $(0)^0 = 0$) be agent i's generalized counting liability index.

Using the corresponding proportional cost-ratio index (4.5), we get the associated cost additive *generalized counting rule*, ϕ^π, given by the payments

$$\phi_i^\pi (M, G, \{\mathcal{D}^j\}_{j \in N}, c) = \sum_{e \in E} \frac{\left(\dfrac{|\overline{\mathcal{D}}^i(e)|}{|\overline{\mathcal{D}}^i|} \right)^\pi}{\sum_M \left(\dfrac{|\overline{\mathcal{D}}^j(e)|}{|\overline{\mathcal{D}}^j|} \right)^\pi} c_e, \tag{4.13}$$

for each agent $i \in M$. In (4.13) the cost of each edge is shared in proportion to the generalized liability index, and the payment of each agent found by summing over the set of edges in the network.

For $\pi = 0$, the rule ϕ^0 divides the cost of edge e equally among all agents for whom e is relevant. If all agents find all edges relevant (i.e., for all agents, all edges are part of some minimal subgraph providing the desired connectivity), the total cost of the network is then split equally. The cost of edges that are irrelevant for everybody is also split equally among the agents (here the convention $(0)^0 = 0$ plays a role).

For $\pi = 1$ the liability index θ^1 is the (canonical) counting index defined in (4.10). The corresponding cost-allocation rule,

$$\phi_i^1(M, G, \{\mathcal{D}^j\}_{j\in N}, c) = \sum_{e\in E} \frac{\dfrac{|\overline{\mathcal{D}}^i(e)|}{|\overline{\mathcal{D}}^i|}}{\sum_M \dfrac{|\overline{\mathcal{D}}^j(e)|}{|\overline{\mathcal{D}}^j|}} c_e, \quad \text{for all } i, \tag{4.14}$$

is dubbed the *counting rule*.

When $\pi \to \infty$, the generalized liability index becomes $\theta^\pi(\mathcal{D}^i, e) = 1$ if $e \in K^i$, and 0 otherwise, so each critical edge is shared equally among the liable agents, and the cost of all other edges are shared equally among all the agents in M. This adapts the equal-need rule (4.4) to the model with redundancy.

Hougaard and Moulin (2014) prove that generalized counting rules are characterized by the axioms of section 4.2.3.

Theorem 4.2 A cost-allocation rule ϕ satisfies the axioms of anonymity, neutrality, consistency, irrelevance of supplementary items, replication, and cost Additivity if and only if it is a generalized counting rule, ϕ^π, given by (4.13) for some $\pi \geq 0$.

Proof [Sketch of argument] We have already noted that the generalized counting rule satisfies all the axioms. Considering the converse claim, we will focus on the impact of irrelevance of supplementary edges. By anonymity and consistency, we have that the cost ratios φ takes the proportional form for a liability index $\theta(\mathcal{D}^i, e)$ meeting properties (4.6) and (4.7). By neutrality and irrelevance of supplementary edges, we can now demonstrate that the liability index takes the form $\theta(\mathcal{D}, e) = g(|\overline{\mathcal{D}}|, |\overline{\mathcal{D}}(e)|)$, where $g(p, q)$ is defined for all integers p, q such that $0 \leq p \leq q \geq 1$, $g(p, q) > 0$ if $q \geq 1$, and $g(p, q) = 0$ if $q = 0$. Indeed, fix e and consider a problem with two agents $i = 1, 2$, letting $p_i = |\overline{\mathcal{D}}^i|$ and $q_i = |\overline{\mathcal{D}}^i(e)|$ and assuming that $q_2 \geq 1$. Let $Y = \{1, 2, \dots\}$ be a set of proxy edges disjoint from E with cardinality $p_1 + p_2$. Now take these proxy edges and add them one by one to the edges in p_1 and p_2, starting with the first subgraph in q_1 and when they have all been added a proxy edge, continuing with subgraphs in $p_1 \setminus q_1$, then starting with q_2, and so forth, ending with the last in $p_2 \setminus q_2$. By irrelevance of supplementary edges, the cost share of edge e is unchanged by these operations. Next, remove one at a time all edges in $E \setminus e$, keeping the proxy edges fixed. By irrelevance of supplementary edges, the cost share of e is unchanged by these operations. So after these operations, we have a problem with edges $Y \cup \{e\}$ and the minimal subgraphs $\overline{\mathcal{T}}^i$ given by

$$\overline{\mathcal{T}}^1 = \{\{e, 1\}, \{e, 2\}, \ldots, \{e, q_1\}, \{q_1 + 1\}, \ldots, \{p_1\}\},$$

$$\overline{\mathcal{T}}^2 = \{\{e, p_1 + 1\}, \ldots, \{e, p_1 + q_2\}, \{q_1 + q_2 + 1\}, \ldots, \{p_1 + p_2\}\}.$$

By the assumption $q_2 \geq 0$, the ratio

$$\frac{\theta(Y \cup \{e\}, \overline{\mathcal{T}}^1, e)}{\theta(Y \cup \{e\}, \overline{\mathcal{T}}^2, e)}$$

is well defined, and by neutrality it only depends on $|Y \cup \{e\}| = 1 + p_1 + p_2$, p_i and q_i for $i = 1, 2$. By construction we have

$$\frac{\theta(Y \cup \{e\}, \overline{\mathcal{T}}^1, e)}{\theta(Y \cup \{e\}, \overline{\mathcal{T}}^2, e)} = \frac{\theta(E, \overline{\mathcal{D}}^1, e)}{\theta(E, \overline{\mathcal{D}}^2, e)} = f(p_1, q_1, p_2, q_2)$$

for some function f. Fixing $p_2 = q_2 = 1$ (that is, all edges are critical to agent 2), we get $\theta(E, \overline{\mathcal{D}}^1, e) = h(|E|) \times f(p_1, q_1)$. Given that the cost ratio takes the proportional form, the term $h(|E|)$ can be dropped.

Finally, it can now be shown that replication implies $\varphi = \varphi^\pi$ for some $\pi \geq 0$. ∎

It can further be shown that adding a standard *unanimity lower bound* condition (stating that the payment of agent i has to exceed $1/m$ of agent i's standalone cost; i.e., $\phi_i \geq \frac{1}{m} \min_{D \in \mathcal{D}^i} c_D$) limits the relevant set of π values in (4.13) to $\pi \in [0, 1]$.

The unanimity lower bound is a natural and compelling condition, since it basically states that a given agent i can never pay less than an equal share of the efficient cost in case all other agents are clones of i.

4.2.5   Limited Reliability

The reason that seemingly redundant connections can have value for network users, and thereby be worth paying for, is the fact that connections often have limited reliability and so may fail. For instance, some connections may be temporarily closed due to maintenance, or simply may be unavailable at random for any number of reasons (e.g., when streaming fails momentarily). So we now consider a situation where users pay for the edges of a network covering their connectivity needs, but the edges may fail. As before, each user has a binary inelastic need that is served if and only if certain subsets of the edges are actually functioning. Again we ask how the total network cost should be divided when individual needs are heterogeneous. Following Hougaard and Moulin (2018), two powerful independence properties can be imposed in addition to cost additivity. The first property, independence of timing, ensures that the cost shares computed ex ante are the expectation, over the random realization of the edges, of shares computed ex post. Cost additivity together with the second condition, dubbed separability across projects, ensure that the cost shares of an edge depend only on the service provided by that edge for a given realization of all other edges. Combining these properties with fair bounds on the liabilities of agents with more or less flexible needs and those of agents for whom an edge is either indispensable or useless, two interesting rules emerge. The ex post rule is the expectation of the equal division of costs among users who end up being served. The needs-priority rule splits the cost first among users for whom an edge is critical ex post, or if there are no such users, among those who end up being served.

Formally, it is straightforward to modify the model to include connections with limited reliability. Suppose that nature chooses the set X, $\emptyset \subseteq X \subseteq E$, of successful (working) edges with probability distribution p. Irrespective of the realization of X, the agents in M still have to cover the total cost for the network c_E: these costs can therefore be considered as sunk costs for the agents. Thus, a *problem with limited reliability* is simply a list $(M, G, \{\mathcal{D}^i\}_{i \in M}, p, c)$.

To include limited reliability in the model has a fundamental impact on the notion of fairness. Here are two main reasons:

• The view on fairness may depend on whether we consider the situation before (ex ante) or after (ex post) the resolution of uncertainties. For instance, ex ante it may seem fair that there is a clear connection between payment and the probability of obtaining connectivity in the network. But ex post we may apply any idea of fairness from the deterministic setting, and this fairness notion may be used ex ante by computing payments in expectation. As we will see below in example 4.7, the ex ante and ex post approaches may result in quite different payments.

• When edges may fail, the relation between flexibility in need and liability in payment becomes ambiguous. Indeed, the more flexible agent now has a bigger chance of obtaining connectivity, speaking in favor of increased liability in payment, even though higher flexibility generally ought to result in lower liability in payment. The question is how these effects ought to be balanced in the choice of allocation rule. The example below will illustrate.

Example 4.7 (Ex ante versus ex post fairness)   Consider the situation where two agents, Ann and Bob, are located at the same node t on a source loop, as illustrated by the graph below.

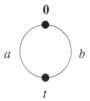

Ann just wants connection to the source 0, while Bob requires 2-connectivity. Thus, Ann may be served by either of the two edges, $\overline{\mathcal{D}}^{Ann} = \{a, b\}$, while Bob needs both, $\overline{\mathcal{D}}^{Bob} = \{ab\}$. Assume that each edge in the loop has a probability of success q. Moreover, assume that p is IID. We thus have the following table.

X	$\emptyset$	a	b	ab
$p(X)$	$(1-q)^2$	$q(1-q)$	$q(1-q)$	q^2

Given the limited reliability of the edges, Ann and Bob have different probabilities of successful connection to the source. In the above case, Ann has ex ante probability $1 - (1-q)^2 = q(2-q)$ of being connected, while Bob's ex ante probability is q^2. A natural idea of fairness from an ex ante perspective could, for instance, entail that the agents should pay in proportion to these probabilities (dubbed the ex ante rule in Hougaard and Moulin, 2018), making Ann pay the share $\varphi^{Ann} = 1 - \frac{q}{2}$ of the total cost (and hence making Bob pay the share $1 - \varphi^{Ann}$). So with very small probability of success (q close to 0), Ann basically pays for everything. Then as $q \to 1$, Ann's share decreases to 0.5.

In contrast, we may view fairness from an ex post perspective once the uncertainty is resolved. If none of the edges worked, neither Ann nor Bob obtained connections, and it seems fair that they split the costs equally. If either a or b worked, only Ann got her desired connection, and it seems fair that she pays for everything. Finally, if both edges worked, both Ann and Bob got their desired connections, and the total cost should be split equally. If payment has to take place ex ante, we can just consider the expected payments (dubbed the ex post rule in Hougaard and Moulin, 2018). In the above case this would result in Ann having the cost share $\varphi^{Ann} = \frac{1}{2}(1-q)^2 + 2q(1-q) + \frac{1}{2}q^2 = 0.5 + q - q^2$. Clearly, this differs from the ex ante approach above: for $q < 0.5$, Ann pays a larger share using the ex ante rule than the ex post rule, while it is the other way around for $q > 0.5$ (for $q = 0.5$, both approaches yield $\varphi^{Ann} = 0.75$).

Finally, notice that Ann is more flexible than Bob, yet using both the ex ante and the ex post rules above, she will always pay weakly more than Bob. As argued, this is not necessarily unfair, since she always has at least as big a chance of getting her desired connection (and often much bigger) than does Bob. Yet, as we approach the deterministic case (as $q \to 1$), it can be argued that Ann should indeed pay less than Bob due to her greater flexibility.

Hougaard and Moulin (2018) argue that a suitable allocation rule ought to make agents indifferent between dividing costs before or after the uncertainties are resolved. This will be the case if the cost shares computed ex ante coincide with the expectation of those computed ex post. This condition defines an axiom dubbed *independence of timing*.

Independence of timing For $p, q \in \Delta(2^E)$ and $\lambda \in [0, 1]$,

$$\phi(\{\mathcal{D}^i\}_{i \in M}, \lambda p + (1 - \lambda)q, c) = \lambda \phi(\{\mathcal{D}^i\}_{i \in M}, p, c)$$
$$+ (1 - \lambda)\phi(\{\mathcal{D}^i\}_{i \in M}, q, c).$$

Cost-allocation rules that are cost additive and satisfy independence of timing will have shares φ for individual edges e of the general form

$$\varphi(e; M, G, \{\mathcal{D}^j\}_{j \in M}, p, c) = \sum_{\emptyset \subseteq X \subseteq E} p(X) \times \varphi^*(e; M, X, \{\mathcal{D}^j\}_{j \in M}, c),$$

$$(4.15)$$

where $\varphi^*(e; M, X, \{\mathcal{D}^j\}_{j \in M}, c)$ allocates the cost of edge e in the deterministic case where only the edges in X are working and available.

Note that the ex ante rule, formally defined as

$$\phi_i^{xa}(M, G, \{\mathcal{D}^j\}_{j \in M}, p, c) = \begin{cases} \dfrac{p(X \in \mathcal{D}^i)}{\sum_{j \in M} p_j(X \in \mathcal{D}^j)} c_E & \text{if } p(\cup_j \mathcal{D}^j) > 0 \\ 1/m & \text{if } p(\cup_j \mathcal{D}^j) = 0 \end{cases},$$

$$(4.16)$$

for every $i \in M$, does not satisfy independence of timing. In contrast, the *ex post rule* does. Formally, the ex post rule is defined as

$$\phi^{xp}(M, G, \{\mathcal{D}^j\}_{j \in M}, p, c) = \left\{ \sum_{X \in 2^E} p(X) \times e[S(X)] \right\} \times c_E, \qquad (4.17)$$

where $S(X) = \{i \in M \mid X \in \mathcal{D}^i\}$ is the set of agents who obtain connectivity given working edges X, and $e[S]$ is the uniform lottery on S, with the convention that $e[\emptyset] = e[N]$.

Note that both the ex ante and the ex post rules assign the same cost share to every edge in E for a given problem. This feature is not compelling, since it ignores a lot of the information provided by the service constraints, revealing the usage of individual edges. Due to the general form of (4.15), it is rather easy to find rules that satisfy both cost additivity and independence of timing, since we can basically choose any deterministic rule and compute cost shares in expectation. Therefore, one such rule is a natural extension of the counting rule (4.14) to the present setup with limited reliability. The definition of the counting rule for deterministic problems in (4.14) is adapted to the probabilistic model by projecting service needs for any realization X as $\mathcal{D}^i(X) = \{D \subseteq X \mid D \subseteq \mathcal{D}^i\}$ and taking expectations over X. Thus, for the *expected counting liability* rule, agent i's cost share of edge e is defined for all $i \in M$ as

$$\varphi_i^{ecl}(e; M, G, \{\mathcal{D}^j\}_{j \in M}, p, c) = \sum_{\emptyset \subseteq X \subseteq E} p(X) \times \frac{\dfrac{|\overline{\mathcal{D}}^i(e, X)|}{|\overline{\mathcal{D}}^i(X)|}}{\sum_{j \in M} \dfrac{|\overline{\mathcal{D}}^j(e, X)|}{|\overline{\mathcal{D}}^j(X)|}}, \quad (4.18)$$

where $\overline{\mathcal{D}}^i(e, X)$ is the set of i's minimal subgraphs containing edge e, given realizations X, with the convention $\frac{0}{0} = \frac{1}{m}$.

Example 4.7 (continued: The expected counting liability rule) Recall the situation from example 4.7 above. Using the expected counting liability rule (4.18), we first need to project the service constraints of each agent onto X. When $X = \emptyset$, no one is served, and Ann and Bob split costs equally, both in case of edge a and edge b. When $X = \{a\}$, Ann pays the full cost c_a when sharing the cost of edge a. Similarly, when $X = \{b\}$, she pays the full cost c_b when sharing the cost of the edge b. But recall that when $X = \{a, b\}$, the counting rule results in shares $(\frac{1}{3}, \frac{2}{3})$ for both edges a and b. Thus, Ann's share using the expected counting liability rule is $\varphi_{Ann}^{ecl} = \frac{1}{2}(1-q)^2 + \frac{1}{2}q(1-q) + q(1-q) + \frac{1}{3}q^2 = \frac{1}{2} + \frac{1}{2}q - \frac{2}{3}q^2$. Now, for $q > \frac{3}{4}$, this makes Ann's share smaller than Bob's, which is indeed reasonable when approaching the deterministic case, because of Ann's higher flexibility.

Hougaard and Moulin (2018) furthermore argue in favor of a second independence property called *separability across edges* (stating that costless deterministic edges should only influence cost shares through their impact on the service constraints). They characterize the class of rules satisfying all three independence properties: cost additivity, independence of timing, and separability across edges. The ex post rule is a member of this class, but the expected counting liability rule is not, because it fails the property of separability across edges.

Within this class, an additional rule is singled out: the *needs priority* rule. This rule assigns specific cost shares for each edge based on the idea that, ex post, agents for whom the edge in question is pivotal (that is, agents served if and only if this edge is working) should share equally, and if no such agents exist, then all agents served should share equally:

$\phi^{np}(M, G, \{\mathcal{D}^j\}_{j \in M}, p, c)$

$$= \sum_{e \in E} \left\{ \sum_{\varnothing \subseteq X \subseteq E} p(X) \times e[S(X) \setminus S(X \setminus e); S(X)] \right\} c_e, \qquad (4.19)$$

with the notation $e[S^1; S^2] = e[S^1]$ if $S^1 \neq \varnothing$, and $e[\varnothing; S^2] = e[S^2]$.

Hougaard and Moulin (2018) provide axiomatic characterizations of both the ex post and the needs priority rules. In particular, a property stating that the more flexible agent should pay most separates the two rules: it is satisfied by the ex post rule ϕ^{xp} but not by the needs priority rule ϕ^{np}. As discussed above, the role of flexibility is ambiguous under limited reliability, which seems to speak in favor of the needs priority rule. Yet both rules are known for their extreme outcomes in deterministic situations, so some kind of weighted average of the two rules may prove more desirable: noting that, the class of rules satisfying cost additivity, independence of timing, and separability across edges is closed by convex combinations.

Example 4.8 (Computing shares of the needs priority rule) Consider the following network consisting of a loop with source 0 (as supplier). Three agents, Ann, Bob, and Carl, all want connection to the source and are located at nodes A, B, and C, respectively, as illustrated below.

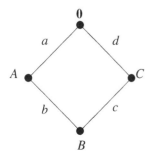

Edges a, b, c, d, are the public goods, all with limited reliability q, and p is IID as in example 4.7. Given connection demands, we get the following (minimal) service constraints for each agent:

$$\overline{\mathcal{D}}^{Ann} = \{a, bcd\}, \ \overline{\mathcal{D}}^{Bob} = \{ab, cd\}, \text{ and } \overline{\mathcal{D}}^{Carl} = \{d, abc\}.$$

In the deterministic cases ($q \in \{0, 1\}$), the three agents should clearly split costs equally, since all have 2-connectivity in the loop. For limited reliability, the situation is more complicated.

Using the ex post rule results in cost shares:

$$\phi^{xp} = \frac{1}{3}(1 + q - 2q^2 + q^3, 1 - 2q + 4q^2 - 2q^3, 1 + q - 2q^2 + q^3).$$

Clearly, Ann and Carl are located symmetrically, so they pay the same amount. In the deterministic cases $q \in \{0, 1\}$, all share equally, as they should. For all other values of q, Bob pays less than Ann and Carl, since he has a lower probability of successful connection to the source.

Using the needs priority rule gives the following cost shares for edges a and b:

$$\phi^{np}(a) = \frac{1}{6}(2 + 2q - q^2 + q^3 - 2q^4, 2 - q - q^2 + q^3 + q^4,$$
$$2 - q + 2q^2 - 2q^3 + q^4),$$

$$\phi^{np}(b) = \frac{1}{6}(2 + 2q - 7q^2 + 12q^3 - 7q^4, 2 - 4q + 11q^2 - 3q^3 - 4q^4,$$
$$2 + 2q - 4q^2 - 9q^3 + 11q^4),$$

and similarly (due to symmetry) for edges c and d.

Let us focus on payment for the edges a and b. For edge a, Ann pays more than Carl, who pays more than Bob for all values of $q \in (0, 1)$. This makes sense, since Ann ought to be more liable than Carl for her direct connection to the source. The picture is more complicated for edge b, where payments vary substantially among agents according to the value of q. For low reliabilities ($q < 0.4$), Ann pays more than Carl, who pays more than Bob (Ann and Carl both have higher probability of getting service than Bob does, and should accordingly pay more). For high reliabilities ($q > 0.6$), Bob pays more than Ann, who pays more than Carl (Bob is now the agent who needs b the most, and accordingly should pay more than both Ann and Carl).

4.2.6 Cost Responsibility

So far, the property of cost additivity has played a central role in the analysis of relevant allocation rules. However, supposing it makes sense to hold individual agents responsible for the cost of individual edges, we will have to abandon the idea of cost additivity and find rules that relate directly to such responsibilities. To motivate this, consider the following example.

Example 4.9 (Cost responsiveness)   The network below illustrates a situation where four nodes $\{a, b, c, d\}$ are connected with edge costs as shown in the graph. Two agents, Ann and Bob, have the following connectivity needs: Ann wants to connect nodes a and c; Bob wants to connect nodes b and d.

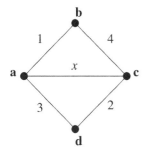

Their service constraints are therefore given by

$$\overline{\mathcal{D}}^{Ann} = \{\{ac\}, \{ab, bc\}, \{ad, cd\}\}$$

and

$$\overline{\mathcal{D}}^{Bob} = \{\{ab, ad\}, \{bc, cd\}, \{ab, ac, cd\}, \{bc, ac, ad\}\}.$$

Using, for instance, the counting rule (4.14), the cost of all edges are shared in proportion $\frac{1}{3} : \frac{1}{2}$, resulting payments being $(\frac{2}{5}(10+x), \frac{3}{5}(10+x))$. As a result, as the cost x of edge ac increases, Ann and Bob keep splitting this cost in the same fixed proportion $\frac{1}{3} : \frac{1}{2}$, even though when x is small ($0 \le x < 1$), both Ann and Bob must use ac in their cost-minimizing connection path, while when x is large ($x > 5$), both of them have cheaper options, excluding the use of edge ac. When both Ann and Bob find the edge ac useful, it seems reasonable that they split its cost equally. When ac becomes irrelevant for both of them, it can be argued that they ought to split its cost equally as well. In between these cases, other ways to share costs can be relevant. Obviously, allocation rules building on this line of reasoning require that we can define a level of cost responsibility of each agent, for each edge, based on their

cost-minimizing connectivity options. But allocation rules that are responsive
to changes in edge costs violate cost additivity, so we will have to abandon
the idea that fair allocation only depends on the way that agents can obtain
service from the network.

A natural way to think about the cost responsibility of a given agent is the
following. Intuitively, agents can only be held responsible for their standalone
cost. That is, for any edge e, which belongs to a cost-minimizing path satisfying agent i's connection demand, agent i is fully responsible for c_e. For the
cost of all other edges, she can only be partly responsible. In particular (and
not unlike the idea behind the irreducible cost matrix from the MCST model
in chapter 2), we can lexicographically lower the maximal edge cost among all
remaining edges until these (truncated) edge costs make a new minimal subgraph of agent i cost efficient. Agent i is then responsible for the truncated
edge cost of the relevant edges in this (these) new efficient path(s). Repeating
this process eventually leads to an assigned cost responsibility for every edge
for agent i.

Formally, let $\gamma(e; \mathcal{D}^i, c)$ be a *cost responsibility function*, assigning to each
edge $e \in E$ a share of the cost c_e, for which agent $i \in M$ is directly responsible.
For every agent i, cost responsibilities depend only on $(M, G, \mathcal{D}^i, c)$, and not
on the service constraints of the other agents.

Denote by $sa_i(\mathcal{D}^i, c) = \min_{D \in \overline{\mathcal{D}}} c_D$ the standalone cost of agent i (i.e., the
lowest cost possible for satisfying i's connection demand), and let $\Gamma(\mathcal{D}^i, c) =$
$\{c' \in \mathbb{R}_+^E \mid c' \leq c$ and $sa_i(\mathcal{D}^i, c') = sa_i(\mathcal{D}^i, c)\}$ be the set of cost functions
bounded above by c, ensuring the same standalone cost.

A natural definition of the cost responsibility function γ_i is as the minimal
element of $\Gamma(\mathcal{D}^i, c)$, that is, $\Gamma(\mathcal{D}^i, \gamma_i) = \gamma_i$. The stepwise procedure below can
be used to determine γ_i for every agent $i \in M$.

The descent algorithm starts at $t = \infty$ with t decreasing until at most 0:

Step 0. Set $\gamma_i(e) = c_e$ for $e \in E^0 = \cup\{D \in \widehat{\mathcal{D}}^i\}$, where $\widehat{\mathcal{D}}^i = \{D \in \overline{\mathcal{D}}^i \mid c_D =$
$sa_i\}$. If $E^0 = E$, stop. Otherwise continue.

Step 1. Let t^1 be the smallest $t \geq 0$ such that $c_{D \cap E^0} + \sum_{D \setminus E^0} \min\{t, c_e\} \geq sa_i$
for all $D \in \overline{\mathcal{D}}^i$. Let E^1 be the union of all $D \in \overline{\mathcal{D}}^i$ for which this inequality is an
equality at t^1 (making $E^0 \subset E^1$). Set $\gamma_i(e) = \min\{t^1, c_e\}$ for $e \in E^1 \setminus E^0$, and
define $c^1 = c$ on E^0 and $c^1 = \min\{t^1, c\}$ on $E^1 \setminus E^0$. Continue if $E^1 \neq E$.

...

Step k. Let t^k be the smallest t, for which $t^{k-1} > t \geq 0$, such that $c_{D \cap E^{k-1}}^{k-1} +$
$\sum_{D \setminus E^{k-1}} \min\{t, c_e\} \geq sa_i$ for all $D \in \overline{\mathcal{D}}^i$. Let E^k be the union of all $D \in \overline{\mathcal{D}}^i$

for which this inequality is an equality at t^k (making $E^{k-1} \subset E^k$). Set $\gamma_i(e) = \min\{t^k, c_e\}$ for $e \in E^k \setminus E^{k-1}$, and define $c^k = c^{k-1}$ on E^{k-1} and $c^k = \min\{t^k, c\}$ on $E^k \setminus E^{k-1}$. Continue if $E^k \neq E$.

The resulting cost responsibility function γ_i of the above descent algorithm minimizes the maximal cost of any edge in Γ:

$$\max_{e \in E} \gamma_i(e) \leq \max_{e \in E} c'_e \text{ for any } c' \in \Gamma(\mathcal{D}^i, c).\tag{4.20}$$

Indeed, if $\max_{e \in E} \gamma_i(e)$ is achieved in E^0, this is trivially true, since $\gamma_i(e) = c'_e$ for such edges. Otherwise the maximum is achieved for some $b \in E^1 \setminus E^0$ such that $c_{D \cap E^0} + \sum_{D \setminus E^0} \min\{t, c_e\} = sa_i$ for some $D \in \overline{\mathcal{D}}^i$. Suppose $c' \in \Gamma(\mathcal{D}^i, c)$ is such that $\max_{e \in E} c'_e < \gamma_i(b) = t^1$. Then we have a contradiction, since $c'_{D \cap E^0} + \sum_{D \setminus E^0} c'_e < sa_i$, because $c' = c$ on $D \cap E^0$, $c'_b < t^1 = \min\{t^1, c_e\}$, and $c'_e \leq \min\{t^1, c_e\}$ for all $e \in D \setminus E^0$.

Example 4.9 (continued: Computing cost responsibilities) To illustrate the descent algorithm above, recall the case in example 4.9, and focus on Bob's cost responsibility when $x = 6$. Bob has four potential (minimal) paths connecting nodes b and d:

$D_1 = \{ab, ad\},$

$D_2 = \{bc, cd\},$

$D_3 = \{ab, ac, cd\},$

$D_4 = \{bc, ac, ad\},$

with respective costs $c_{D_1} = 4$, $c_{D_2} = 6$, $c_{D_3} = 9$, $c_{D_4} = 13$. So Bob's standalone cost is 4. Thus, $E^0 = \{ab, ad\}$, and his cost responsibility for the edges ab and ad are equal to their full cost: $\gamma_{Bob}(ab) = 1$, $\gamma_{Bob}(ad) = 3$.

Moving on to step 1 in the algorithm, we see that the four inequalities become

$D_1 : c_{D_1} = 4 \geq 4 = sa_{Bob},$

$D_2 : \min\{t, 4\} + \min\{t, 2\} \geq 4,$

$D_3 : \min\{t, 6\} + \min\{t, 2\} \geq 4 - 1 = 3,$

$D_4 : \min\{t, 4\} + \min\{t, 6\} \geq 4 - 3 = 1.$

So we have $t^1 = 2$, with the first two inequalities binding. Thus, $E^1 = \{ab, ad, bc, cd\}$, and the cost responsibilities for the edges bc and cd become $\gamma_{Bob}(bc) = 2$, $\gamma_{Bob}(cd) = 2$.

Since $E^1 \neq E$, we move on to step 2. The four inequalities $\sum_{D \setminus E^1} \min\{t, c_e\} \geq 4 - c^1_{D \cap E^1}$ are now given by

$D_1 : c_{D_1} = 4 \geq 4 = sa_{Bob}$,

$D_2 : c^1_{D_2} = 4 \geq 4 = sa_{Bob}$,

$D_3 : \min\{t, 6\} \geq 4 - 3 = 1$,

$D_4 : \min\{t, 6\} \geq 4 - 5 = -1$.

So we get $t^2 = 1$ and $E^2 = E$. We are therefore done and now have the final cost responsibility for edge ac: $\gamma_{Bob}(ac) = 1$.

Similarly, we can show that Ann has cost responsibilities

$\gamma_{Ann}(ab) = 1$, $\gamma_{Ann}(bc) = 4$, $\gamma_{Ann}(ad) = 3$, $\gamma_{Ann}(cd) = 2$, and $\gamma_{Ann}(ac) = 5$.

Once cost responsibilities are determined for every agent, we can assign the cost of every edge in a serial fashion based on these responsibilities. Define the *serial responsibility* rule as follows.

• Fix an edge e, and label agents according to increasing cost responsibility:

$0 \leq \gamma_1(e) \leq \gamma_2(e) \leq \cdots \leq \gamma_m(e) \leq c_e$.

• Then the payment of agent $i \in M$, for edge $e \in E$, is determined by

$$\frac{1}{m}\gamma_1(e) + \frac{1}{m-1}(\gamma_2(e) - \gamma_1(e)) + \cdots + \frac{1}{m-i+1}(\gamma_i(e) - \gamma_{i-1}(e))$$
$$+ \frac{1}{m}(c_e - \gamma_m(e)). \tag{4.21}$$

• For every agent, add up the cost shares over the set of edges to get each agent's total payment.

The serial responsibility rule satisfies all the axioms of theorem 4.2 except cost additivity and consistency.

Example 4.9 (continued: Using the serial responsibility rule) Computing the cost shares of the serial responsibility rule (4.21) on data from example 4.9 (with $x = 6$), we get the following.

The total cost of the network is 16. The cost responsibilities of individual edges are

$ab : \gamma_{Ann}(ab) = 1 = \gamma_{Bob}(ab) = c_{ab}$,

$ac : \gamma_{Bob}(ac) = 1 < 5 = \gamma_{Ann}(ac) < 6 = c_{ac}$,

$ad : \gamma_{Ann}(ad) = 3 = \gamma_{Bob}(ad) = c_{ad}$,

$bc: \gamma_{Bob}(bc) = 2 < 4 = \gamma_{Ann}(bc) = c_{bc}$,

$cd: \gamma_{Ann}(cd) = 2 = \gamma_{Bob}(cd) = c_{cd}$.

Thus, the cost of edge ab is split equally. For edge ac, they split the cost equally up to 1, Ann covers the remaining amount up to 5, and the residual of 1 (for which neither of them is responsible) is split equally as well. For edge ad, they split the cost equally. For edge bc, they split the cost equally up to 2, and Ann covers the remaining amount of 2. Finally, for edge cd, they split the cost equally.

The resulting payment is therefore 11 for Ann and 5 for Bob. Compared to the counting rule there is a radical difference: applying (4.14) gives payments of 6.4 and 9.6 for Ann and Bob, respectively. Using the counting rule, Bob is made more liable than Ann, because relatively speaking, each edge is more central for Bob's connectivity options than it is for Ann's. However, in the sense of getting the cheapest connection, the edges matter less for Bob than they do for Ann, thus making Ann more liable in terms of cost responsibility.

Example 4.10 (Comparing the serial responsibility and counting rules) We can generalize the findings from example 4.9. Fix a set of nodes N ($|N| = n$), and consider the complete graph. Each agent wants to connect two specific nodes. Normalize all edge costs to 1.

Say agent $i \in M$ wants to connect $s, s' \in N$. Applying the counting rule, the cost-ratio indices for agent i, φ_i, are computed as follows. There are $(n - 2)$ $(n - 3) \cdots (n - k)$ paths of length k connecting s to s' (without repetition of any nodes). Therefore,

$$|\overline{\mathcal{D}}^i| = 1 + (n - 2) + (n - 2)(n - 3) + \cdots + (n - 2)(n - 3) \cdots 2 \times 1$$
$$= f(n - 2),$$

where

$$f(k) = k! \left(2 + \frac{1}{2!} + \frac{1}{3!} + \cdots + \frac{1}{k!} \right).$$

Note that $(2.66)k! \leq f(k) \leq (2.75)k!$ for all $k \geq 3$. Next we have $|\overline{\mathcal{D}}^i(ss')| = 1$, $|\overline{\mathcal{D}}^i(sz)| = f(n - 3)$, and

$$|\overline{\mathcal{D}}^i(zz')| = 2\{f(n - 4) + (n - 4)f(n - 5) + (n - 4)(n - 5)f(n - 6)$$
$$+ \cdots + (n - 4)! \times 2 + (n - 4)!\}$$

$$\Rightarrow 2((2.66)(n - 6) + 8) \leq \frac{|\overline{\mathcal{D}}^i(zz')|}{(n - 4)!} \leq 2((2.75)(n - 6) + 8)$$

for nodes $z, z' \in N$ distinct from s and s'. So for large enough n, $|\overline{\mathcal{D}}^i(sz)| \simeq$ $(2.7)(n-3)!$, and $|\overline{\mathcal{D}}^i(zz')| \simeq (5.4)(n-3)!$.

We conclude that $\varphi_i(ss') \ll \varphi_i(sz) < \varphi_i(zz')$ for any $z, z' \in N$ different from s and s'.

Consequently, if the graph is large, the counting rule yields an unpalatable division of costs when Ann wants to connect ss', Bob wants to connect ff', and these four nodes are distinct: Ann pays essentially nothing for ss' compared to Bob, she pays half as much as Bob for sz ($z \neq f, f'$), and they pay equal shares for edges like sf.

In contrast, the serial responsibility rule captures the fact that ss' is more important for Ann than any other edge in terms of minimizing her connection cost. We have $E^0 = \{ss'\}$. In the descent algorithm, the first critical point is $t^1 = \frac{1}{2}$, corresponding to all minimal subgraphs $D \in \overline{\mathcal{D}}^{Ann}$ of the form $D = \{sz, zs'\}$, so $E^1 = \{ss', sz, zs', \text{ for all } z \neq s, s'\}$. The next step is $t^2 = 0$, because any other minimal subgraph contains exactly two edges in E^1. Therefore,

$$\gamma_A(ss') = 1, \quad \gamma_A(sz) = \frac{1}{2}, \quad \gamma_A(zz') = 0$$

for all $z, z' \neq s, s'$.

So when Ann wants to connect ss', Bob wants to connect ff', and these four nodes are distinct, then Ann pays the full cost of $ss', sz, s'z$, for all $z \neq s$, s', f, f', Bob pays the full cost of $ff', fz, f'z$, for all $z \neq s, s', f, f'$, and the cost of any other edge is split equally.

4.3   Capacity Networks

We now turn to network design problems. Cost information for the complete graph is available, and we need to determine the optimal network and subsequently share its total cost among the users.

One such design problem is introduced in Bogomolnaia et al. (2010) as a *capacity problem*: a set of agents share a network, allowing them to communicate or distribute some goods. The traffic between agents i and j requires a certain capacity, and capacity is correlated with cost. We want to find a feasible network (satisfying all capacity requirements) with minimal total cost and allocate this cost fairly among network users.

In many ways the model resembles the MCST model from chapter 2, but there are a few important differences. The network has no root, and there are externalities in the sense that the cost of connecting any pair of agents depends on the capacity requirements of other agents as well as on the chosen spanning tree.

4.3.1 The Model

The set of agents is identified by the set of nodes $N = \{1, \ldots, n\}$ in the complete graph. Assume that connecting any pair of agents $(i, j) \in 2^N$ requires a given capacity $k_{ij} \geq 0$. Let K be the $n \times n$ capacity matrix (setting $k_{ii} = 0$). Moreover, assume that the cost of constructing α units of capacity is equal to α dollars, independent of which connection we consider. This last assumption is somewhat simplistic, since in practice it can easily be imagined that the cost of delivering capacity k between agents i and j differs from the cost of delivering capacity k between agents s and t.

A *capacity network* on N is a graph $G = (N, E)$ with weights (capacities) c_e for each edge $e \in E$. A capacity network is *feasible* if it can accommodate all capacity requirements given by the capacity matrix K. That is, for any two agents $i, j \in N$, there exists at least one connected path in G where the capacity of each edge in the path is at least k_{ij}. A *minimum-cost capacity network* is a feasible capacity network that minimizes the total cost $\sum_{e \in E(G)} c_e$ among all feasible networks.

It is intuitively clear that a minimum-cost capacity network must be a spanning tree $\mathcal{T}$: indeed, in any cycle, the cheapest (lowest capacity) link will be redundant. It is well known that a minimum-cost capacity network $\mathcal{T}^*$ can be determined by computing the *maximum*-cost spanning tree. Since it is not a rooted tree, it is natural to use the Kruskal algorithm, ordering edges by decreasing capacity cost (see chapter 1).

Denote by (N, K) a *capacity problem*, and denote by $\mathcal{T}^{\max}(N, K)$ the associated set of minimum-cost capacity networks. Let $\kappa(K) = \sum_{e \in \mathcal{T}^*} c_e$ be the total cost of $\mathcal{T}^* \in \mathcal{T}^{\max}(N, K)$.

Let ϕ be a cost-allocation rule defined on the set of all capacity problems. That is, for each problem (N, K), ϕ associates an n-dimensional vector of cost shares $\phi_i(N, K)$ satisfying budget balance: $\sum_{i \in N} \phi_i(N, K) = \kappa(K)$.

Example 4.11 (Computing a minimum-cost capacity network) Consider four agents, $N = \{1, 2, 3, 4\}$, with capacity requirements

$$(k_{12}, k_{13}, k_{14}, k_{23}, k_{24}, k_{34}) = (1, 2, 3, 4, 5, 6).$$

The problem is to connect all four agents while ensuring that the capacity requirements are satisfied for every pair of agents, by all edges connecting them. As mentioned above, this is done by computing the maximum-cost spanning tree using the Kruskal algorithm. Straightforward application of the Kruskal algorithm gives the unique minimum-cost capacity network $\mathcal{T}^* = (14, 24, 34)$ with total cost $3 + 5 + 6 = 14$. (Indeed, for any numbers $k_{34} > k_{24} > k_{23} > k_{14} > k_{13} > k_{12}$, $\mathcal{T}^* = (14, 24, 34)$ will remain the unique minimum-cost capacity network). Notice that if any edge in the maximum-cost spanning tree is replaced by another, less costly edge, some (pairwise) capacity requirement is violated by the resulting network. Note also that the capacities are public goods in the sense that several agents can use the same edge and its capacity to obtain the desired connectivity without rivalry.

4.3.2 Allocation Rules

Since there is no root in the optimal network, the following cost-allocation rule seems rather natural.

For a given minimum-cost capacity network $\mathcal{T}^* \in \mathcal{T}^{\max}(N, K)$, the cost of every edge $ij \in \mathcal{T}^*$ is shared equally between the adjacent nodes (agents) i and j. Thus, every agent $i \in N$ pays

$$y_i^{\mathcal{T}^*, \frac{1}{2}} = \frac{1}{2} \sum_{j: ij \in \mathcal{T}^*} k_{ij}. \tag{4.22}$$

The *Bird$^{\frac{1}{2}}$ rule*, $\phi^{B\frac{1}{2}}$, is defined as the uniform average of these $\mathcal{T}^*$-specific allocations over the set of all minimum-cost capacity networks of the problem (N, K):

$$\phi^{B\frac{1}{2}} = \frac{1}{|\mathcal{T}^{\max}|} \sum_{\mathcal{T}^* \in \mathcal{T}^{\max}} y^{\mathcal{T}^*, \frac{1}{2}}. \tag{4.23}$$

Another straightforward approach to allocate $\kappa(K)$ comes from selecting an artificial root and using the Bird rule, given the selected root, as follows. Letting a given agent play the part of a nonpaying "root," with probability $1/n$, the remaining $n - 1$ agents pay according to the usual Bird rule (2.17). That is, they pay the cost of their adjacent upstream edge (given the choice of root) in the resulting rooted fixed tree. Averaging over all agents, and over all minimum cost capacity networks, we obtain the *minimum-cost capacity Bird rule* ϕ^B.

The same construction can be used to derive rules based on any allocation scheme for rooted fixed trees (as analyzed in section 2.1.2), for instance, we could take the serial approach. Let a given agent play the part of a nonpaying "root." With probability $1/n$, the remaining $n - 1$ agents pay according to the usual serial scheme (2.16). Averaging over all agents, and over all minimum-cost capacity networks, we obtain the *minimum-cost capacity serial rule* ϕ^S.

Example 4.11 (continued: Computing variations of Bird and serial rules)
Recall the case in example 4.11 above, involving four agents $\{1, 2, 3, 4\}$ with capacity requirements $(k_{12}, k_{13}, k_{14}, k_{23}, k_{24}, k_{34}) = (1, 2, 3, 4, 5, 6)$, and the

(unique) minimum-cost capacity network $\mathcal{T}^* = (14, 24, 34)$ with a minimal cost of 14.

Thus, the Bird$^{\frac{1}{2}}$ rule gives

$$\phi^{B^{\frac{1}{2}}} = (1.5, 2.5, 3, 7).$$

Here, agent 4 pays half of the total amount due to his central position in the optimal network. Thus, in effect, the Bird$^{\frac{1}{2}}$ rule punishes hubs in the network.

Using the minimum-cost capacity Bird rule, we get allocations $(0, 5, 6, 3)$ if agent 1 is selected as the nonpaying root, $(3, 0, 6, 5)$ if agent 2 is the root, $(3, 5, 0, 6)$ if agent 3 is the root, and $(3, 5, 6, 0)$ if agent 4 is the root (see figures below).

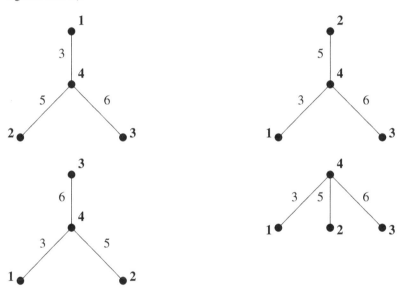

Consequently,

$$\phi^B = \left(\frac{1}{4}9, \frac{1}{4}15, \frac{1}{4}18, \frac{1}{4}14\right) = (2.25, 3.75, 4.5, 3.5).$$

Here, agent 4 pays considerably less than in the Bird$^{\frac{1}{2}}$ solution. Likewise, using the serial rule gives

$$\phi^S = \left(\frac{1}{4}12.67, \frac{1}{4}18, \frac{1}{4}20.67, \frac{1}{4}4.67\right) = (3.16, 4.5, 5.16, 1.16).$$

Here, agent 4, being the hub, pays even less.

4.3.3   The Capacity Problem as a Game

A capacity problem can also be modeled as a cooperative game. The value of any coalition $S \subseteq N$ is naturally given as the minimal cost of connecting all agents in S when any pair of agents in the complement, $N \setminus S$, requires zero capacity. That is, the value is the efficient cost of connecting all agents in S when only capacity requirements from agents in S count. Indeed, members in S should not be held responsible for capacity requirements among members in $N \setminus S$.

Formally, for any coalition $S \subseteq N$, let $K^{|S}$ be the capacity matrix K restricted to the capacity needs for coalition S, that is, $k_{ij}^{|S} = 0$ if $ij \in 2^{N \setminus S}$, and $k_{ij}^{|S} = k_{ij}$ otherwise. So the standalone cost of coalition S is given by

$$c(S, K) = \kappa(K^{|S}). \tag{4.24}$$

It can be shown that the cost function (4.24) is concave so, for instance, the Shapley value of the game (N, c), defined by (4.24), satisfies the standalone core conditions. In fact, so does both the *minimum-cost capacity* Bird rule and the Bird$^{\frac{1}{2}}$ rule, while the *minimum-cost capacity serial* rule does not. (Indeed: consider a three-agent example, where $(k_{12}, k_{13}, k_{23}) = (1, 0, 4)$. The derived game (4.24) consequently becomes $c(1) = 1$, $c(3) = 4$, and $c(S) = 5$, otherwise. Here $\phi^S = (1.33, 0.83, 2.83)$, where $\phi_1 = 1.33 > 1 = c(1)$.)

Example 4.11 (continued: The capacity game)   Recall the complete capacity graph of example 4.11 above. The capacity game defined by (4.24) is given by costs

$$c(1) = 6,\ c(2) = 10,\ c(3) = 12,\ c(4) = 14,$$

$$c(12) = 12,\ c(13) = 13,\ c(14) = 14,\ c(23) = 13,\ c(24) = c(34) = 14,$$

$$c(123) = c(124) = c(134) = c(234) = 14,$$

$$c(1234) = 14.$$

The core is rather large in this case, so even the serial rule ϕ^S results in a core allocation. As mentioned, the Shapley value of the game defined by (4.24), given by the allocation $x^{Sh} = (1.83, 3.17, 4, 5)$, is a core selection, since the game is concave. Compared to the Bird and serial allocations, the result of using the Shapley value is more in line with the Bird$^{\frac{1}{2}}$ rule, where agent 1 pays least, and agent 4 most (simply because of the standalone costs are being smallest respectively largest for these two agents).

Although both Bird-like rules result in core allocations, they suffer from other shortcommings (basically hangovers from the MCST model). As shown in Bogomolnaia et al. (2010), neither rule is continuous in capacity costs k (which implies that small changes in costs may lead to large jumps in cost shares). In addition, the minimum-cost capacity Bird rule ϕ^B further violates the standard cost-monotonicity requirement, stating that cost shares are weakly increasing in capacity costs k_{ij} for all i, j. Somewhat surprisingly, the Shapley value of the game defined by (4.24) also violates cost monotonicity.[4]

In a similar fashion as in the MCST model from chapter 2, we can define the *irreducible capacity matrix* as the largest matrix K^*, weakly above K, with the same optimal cost as K. That is, $k_{ij}^* = \min_{e \in p(i,j)} k_e$, where $p(i, j)$ is the path connecting i and j in a minimum-cost capacity network $\mathcal{T}^*$.

Mimicking the idea of the Folk solution for the MCST problem defined in (2.32), we can solve the allocation problem by using the Shapley value of the game defined in (4.24), but this time with respect to the irreducible capacity matrix K^*. It can be shown that this solution produces a core allocation and also satisfies cost monotonicity and continuity.

Example 4.11 (continued: A game based on the irreducible capacity matrix)
Recall the problem with four agents $\{1, 2, 3, 4\}$ and capacity requirements $(k_{12}, k_{13}, k_{14}, k_{23}, k_{24}, k_{34}) = (1, 2, 3, 4, 5, 6)$, unique minimum-cost capacity network $\mathcal{T}^* = (14, 24, 34)$, and minimal cost 14. Clearly, the irreducible capacities become $(k_{12}^*, k_{13}^*, k_{14}^*, k_{23}^*, k_{24}^*, k_{34}^*) = (3, 3, 3, 5, 5, 6)$. So the game defined in (4.24) with respect to K^* becomes

$$c^*(1) = 9, \, c^*(2) = 13, \, c^*(3) = 14,$$

$$c^*(12) = 13, \, c^*(13) = 14, \, c^*(23) = 14,$$

and $c^*(S) = c(S)$ otherwise.

The Shapley value with respect to c^* is given by the allocation

$$x^{Sh*} = (2.25, 3.58, 4.08, 4.08).$$

Note that this allocation is more equally distributed than the Shapley value of the original game, $x^{Sh} = (1.83, 3.17, 4, 5)$.

---

4. As shown by the example in Bogomolnaia et al. (2010): let K be defined by $k_{12} = 1$, and $k_{ij} = 1$ for all $i \in \{2, 3\}$ and all $j \in \{4, 5, 6, 7, 8\}$, and $k_{..} = 0$ otherwise. Let K' be defined as K except for $k'_{13} = 1$. So agent 1's capacity requirements have weakly increased going from K to K', but $\phi_1^{Sh}(N, K) = 0.5 > \frac{83}{168} = \phi_1^{Sh}(N, K')$.

Bogomolnaia et al. (2010) show that the Shapley value with respect to (4.24) for the irreducible capacity matrix is a member of a broader family of rules (the so-called piecewise linear solutions) that satisfy cost monotonicity and are continuous core selections.

So adding externalities to the network design problem, in the form of pairwise capacity requirements, is relatively straightforward to handle (at least given the simplifying assumptions of the present model). We can still use variations of well-known allocation rules, and the naturally associated cooperative game is concave. So there are many core stable allocation rules to choose among, including a large class satisfying further desirable properties, like cost monotonicity and continuity.

4.4   Minimum-Cost Connection Networks

Continuing with network design problems, let us now discuss a direct generalization of the MCST model, assuming that agents have individual connection demands in the form of two specific nodes (locations) that they want connected. Agents do not necessarily want to connect to the same node, so there is no root, and agents are therefore no longer identified by the set of nodes. Otherwise the model assumptions mimic those of the MCST model.

Generalizing connection demands has a radical impact. For instance, the derived cooperative game is no longer guaranteed to be balanced, so we may end up in situations where no cost allocation is stable in the sense of the core. Moreover, imposing two relevant independence properties seriously limits the range of potential allocation rules, in contrast to the case of the original MCST model.

4.4.1   The Model

Let $M = \{1, \dots, m\}$ denote the (finite) set of agents, and let $\mathcal{L}$ denote a set of potential nodes. Agents' connection demands are given by pairs of nodes that they want connected either directly or indirectly. That is, every agent $i \in M$ has a demand $(a_i, b_i) \in \mathcal{L} \times \mathcal{L}$. Assume that $|\mathcal{L}| \geq 2m + 1$. We say that the profile of demands $P = (a_i, b_i)_{i \in M}$ is a *connection structure*, and denote by $\mathcal{P}$ the set of such connection structures.

The *cost structure* C describes the costs of all edges in the complete graph $G(N_C, E^{N_C})$ with $N_C \subseteq \mathcal{L}$, where $c_{jj} = 0$ for all $j \in N_C$, and $c_{jk} > 0$ for all $j, k \in N_C$ where $j \neq k$. Since the network is undirected, $c_{jk} = c_{kj}$. Furthermore, edges are assumed to be public goods.

A *connection problem* consists of a connection structure and a cost structure (P, C). A *connection network* related to a connection problem (P, C) is a graph $G(N_C, E)$ satisfying all connection demands in P. Let $v(G, C) = \sum_{jk \in G} c_{jk}$ be the total cost of the connection network G. A *minimum-cost connection network* (MCCN) for connection problem (P, C) is a connection network minimizing the total cost. That is, G is MCCN if and only if $v(G, C) \leq v(H, C)$ for every connection network H. Denote by $\mathcal{M}(P, C)$ the set MCCNs for connection problem (P, C). Clearly every MCCN is either a tree or a forest. Thus, for every agent $i \in M$, there is a unique path $p_i(G, P)$ connecting i's nodes (a_i, b_i).

A *cost-allocation problem* (G, P, C) consists of a connection problem (P, C) and an associated MCCN $G \in \mathcal{M}(P, C)$. Let $\mathcal{U}$ be the set of cost-allocation problems, and let $\mathcal{U}_U \subseteq \mathcal{U}$ be the set of cost-allocation problems with undemanded nodes, that is, $(G, P, C) \in \mathcal{U}_U$ if $N_C \setminus \cup_{i \in M}(a_i, b_i) \neq \emptyset$.

In the present context, a *cost-allocation rule* $\phi: \mathcal{U} \to \mathbb{R}_+^m$ maps a cost-allocation problem to a vector of (nonnegative) payments satisfying budget balance, that is, $\sum_{i \in M} \phi_i(G, P, C) = v(G, C)$.

Example 4.12 (A cooperative MCCN game) As in the MCST model, we can associate a cooperative game (M, u) with each connection problem (P, C). For a given coalition of agents $S \subseteq M$, let $P_{|S}$ be the connection demands of the agents in S. For every $S \subseteq M$ we therefore have a subproblem $(P_{|S}, C)$, for which there exists a set of minimum-cost connection networks satisfying the demands in $P_{|S}$, with total cost $v(G_{|S}, C)$. Setting $u(S) = v(G_{|S}, C)$, for all $S \subseteq M$, defines the game. Various solution concepts, like the Shapley value of (M, u), can now be used.

However, since the MCCN model is a generalization of the MCST model, we know that if the problem includes nodes that are not demanded by anybody (equivalent to the case of Steiner nodes in the MCST model), the core of the derived cooperative game may be empty (see example 2.11).

4.4.2 Axioms

Following Hougaard and Tvede (2015), let us now examine two independence properties that have a rather striking impact on the choice of allocation rule.

The first property states that if two cost-allocation problems involve the same set of nodes and connection demands, and have the same efficient network (MCCN), where the costs of any edge in the MCCN are identical, then agents should pay the same irrespective of cost differences on edges outside the common MCCN.

Unobserved information independence For all cost-allocation problems $(G, P, C), (G, P, D) \in \mathcal{U}$ with $N_C = N_D$ and $c_{jk} = d_{jk}$ for all $jk \in G$, $\phi(G, P, C) = \phi(G, P, D)$.

A stronger version of this axiom allows the number of nodes to differ between problems.

Strong unobserved information independence   For all cost-allocation problems $(G, P, C), (G, P, D) \in \mathcal{U}$ with $c_{jk} = d_{jk}$ for all $jk \in G$, $\phi(G, P, C) = \phi(G, P, D)$.

The relevance of (strong) unobserved information independence is particularly clear in an implementation context (as we will see in chapter 6, where we revisit the MCCN model). If payments depend on the cost of edges that are not used in the resulting efficient network, then agents will have conflicting interests over the size of costs that are never realized and therefore counterfactual. But notice that if we represent the cost-allocation problem as a cooperative game (as defined in example 4.12 above), the coalitional values of this game will depend on the cost of edges outside the efficient network. Thus, game-theoretic solution concepts applied to such a game violate unobserved information dependence (e.g., if we apply the Shapley value).

The second independence property states that if a connection problem has multiple MCCNs, agents' resulting payments should not depend on which of those MCCNs is implemented.

Network independence   For all cost-allocation problems $(G, P, C), (G', P, C) \in \mathcal{U}, \phi(G, P, C) = \phi(G', P, C)$.

The relevance of network independence is also particularly clear in the context of implementation. If payments depend on the specific structure of the realized network, then agents may hold conflicting views on which efficient network to implement, and this conflict becomes an obstacle for implementing an efficient network in the first place (as will be shown in chapter 6). If allocation rules are based on the particular configuration of the network, then network independence will be violated. For instance, network independence is violated by the equal need rule, defined in (4.4), which splits the cost of each edge in the MCCN equally among agents with a path p_i using this edge and adds up over edges.

4.4.3   Characterization Results

Hougaard and Tvede (2015) show that unobserved information independence and network independence jointly characterize the class of allocation rules where payments only depend on connection demands, the set of nodes, and the total cost of the efficient network.

Theorem 4.3 An allocation rule ϕ on $\mathcal{U}_U$ satisfies unobserved information independence and network independence if and only if

$$\phi(G, P, C) = \gamma(P, N_C, v(G, C)), \tag{4.25}$$

where γ maps tuples $(P, N_C, v(G, C))$ to $\mathbb{R}_+^m$.

Proof [Sketch of argument] First, we observe that if there is an unused node (here u), costs can be transferred between edges in a given MCCN without affecting the allocation of payments, invoking the two independence properties. For instance, consider the case where Ann wants to connect nodes r and s, Bob wants to connect nodes s and t, and the MCCN is given by $G = \{rs, st\}$ in panel (a) in the figure below. By unobserved information independence, the cost allocation must be the same in panels (a) and (b). Note that in panel (b), the graph $G' = \{ru, su, tu\}$ is MCCN as well, and by network independence the cost allocation must be the same whether G or G' is chosen. Say we choose G'; then by unobserved information independence, the costs of edges rs and st can be changed as in panel (c) without changing the cost allocation. Eventually we can end up as in panel (d) with the same cost allocation from (a) to (d).

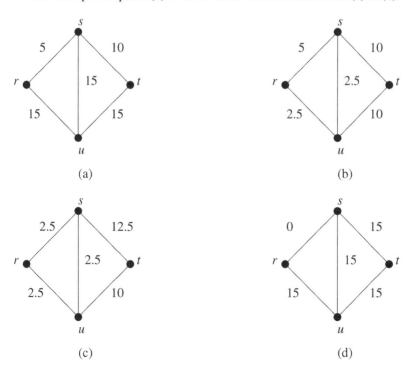

For a general problem (G, P, C), we do the following:

- free an undemanded node;
- add the edge $a_i b_i$ to G for some $i \in M$; and
- remove the path from a_i to b_i in G, and let $c_{a_i b_i}$ be equal to the sum of the costs of the removed edges.

Repeat these steps for all $i \in M$ such that G is transformed into a MCCN with direct connection between connection demands, and move costs to an arbitrary distribution in this network using the observation above.

Next, do the same for another problem (G', P, C'), for which $v(G, C) = v(G', C')$. By the two independence properties and the observation above, payments will be the same in both problems (see also example 4.13). ∎

Example 4.13 (Invoking the two independence properties) Consider the two allocation problems: (G, P, C) in panel (a), and (G', P, C') in panel (b) in the figure below, with MCCNs $G = \{ru, su, tu\}$ and $G' = \{rt, st\}$, and two agents, Ann and Bob, wanting to connect r, s and t, s, respectively, in both problems. Note that $v(G, C) = v(G', C') = 20$.

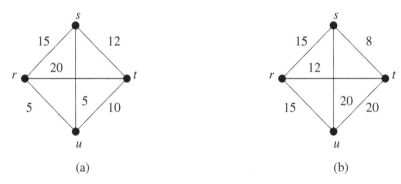

(a)   (b)

We can change from the allocation problem in panel (a) to the allocation problem in panel (b) through a series of transformations based on repeated application of unobserved information independence and network independence, which leaves the cost allocation unchanged.

First, by unobserved information independence, we are free to change the cost of edges not in G, say, to $\widehat{C}$, as shown in panel (a1) in the figure below, without changing payments (i.e., $\phi(G, P, C) = \phi(G, P, \widehat{C})$). Note that now the graph $H = \{rs, st\}$ is a also MCCN in the connection problem $(P, \widehat{C})$ of panel (a1). To free node u, we therefore consider H. By network independence, $\phi(G, P, \widehat{C}) = \phi(H, P, \widehat{C})$. By unobserved information independence, we are free to change the cost of edges not in H, say, to $\widetilde{C}$, as shown in panel (a2), without changing payments (i.e., $\phi(H, P, \widehat{C}) = \phi(H, P, \widetilde{C})$).

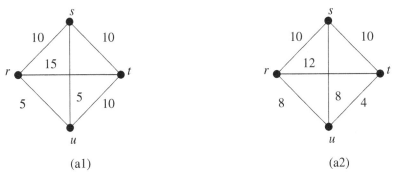

(a1)   (a2)

Next, we can use the fact that G is MCCN in the connection problem $(P, \widetilde{C})$ of panel (a2) and network independence to get $\phi(H, P, \widetilde{C}) = \phi(G, P, \widetilde{C})$. By unobserved information independence, we can change the cost of edges not in G, say, to $\check{C}$ as in panel (a3) in the figure below, without changing payments (i.e., $\phi(G, P, \widetilde{C}) = \phi(G, P, \check{C})$). Finally, since G' is MCCN in the connection problem $(P, \check{C})$ of panel (a3), we can apply network independence to get $\phi(G, P, \check{C}) = \phi(G', P, \check{C})$. And by unobserved information independence, we can set cost of edges not in G' as in panel (a4) without changing payments.

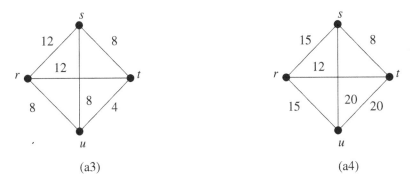

(a3)   (a4)

Noting that the connection problem in panel (a4) is identical to (P, C'), and the allocation problem (G', P, C') is the problem in panel (b), we are done.

The next example compares the result of theorem 4.3 to the case of the MCST model.

Example 4.14 (The MCST model)   The MCST model is a special case of the MCCN model where $N_C = \{0, 1, \ldots, m\}$ and $(a_i, b_i) = (0, i)$ for all $i \in M$. In the MCST model, allocation rules related to cooperative games based on the irreducible cost matrix, like the Folk solution (2.32), satisfy unobserved information independence, because in the problem represented by the irreducible cost matrix, all edges in the complete graph may appear in an efficient network (note that in the general MCCN model, we cannot define an irreducible cost matrix). These solutions also satisfy network independence, so in the MCST model, the class of allocation rules satisfying unobserved information independence and network independence is much richer.

If we replace unobserved information independence with *strong* unobserved information independence in theorem 4.3 above, we get the class of allocation rules where payments only depend on connection demands P and the total cost of the efficient network $v(G, C)$.

Moreover, adding *scale covariance* (which as usual means that payments are homogeneous of degree one in connection costs) to the two independence properties from theorem 4.3 implies that allocation rules take the form

$$\phi(G, P, C) = \delta(P, N_C)v(G, C). \tag{4.26}$$

Again, replacing unobserved information independence with *strong* unobserved information independence implies that payments only depend on connection demand; that is, they take the form

$$\phi(G, P, C) = \rho(P)v(G, C). \tag{4.27}$$

So invoking the two independence properties leads to violations of individual rationality. The full implication of this observation becomes clear in chapter 6, where we return to the MCCN model in the context of implementing an efficient network.

4.5   Network (Value) Games

Jackson and Wolinsky (1996) introduce a model where societal network value depends directly on the specific network structure that emerges when agents interact with one another, for instance, when trading, cooperating, or communicating. So agents have full discretion to form—or cut—connections to other agents. This is a very different situation from what we have considered so far, where networks were either fixed or resulted from a centrally coordinated network design problem.

This way of modeling network formation has many interesting economic aspects (see, e.g., the book Jackson, 2008). The specific problem that we address in the present section is how to allocate the total value of a network resulting from such an interaction-based network formation process. The model is richer than the standard TU game, since the same group of agents may be connected in difference network configurations, and these configurations may have different values for the group. Thus, we examine if, and how, standard allocation rules from cooperative game theory can be extended to this new setup, dubbed a *network game*.

4.5.1 The Model

Let agents be identified by nodes, and fix a finite set of agents $N = \{1, \ldots, n\}$ connected in some undirected network structure. Let $\mathcal{G}$ be the set of all possible undirected network structures on N, that is, the set of all subgraphs of the complete (simple) graph G^N.

For any graph $G \in \mathcal{G}$, the total value of G is given by a function $v : \mathcal{G} \to \mathbb{R}$ (normalized such that $v(G) = 0$ for the completely disconnected graph consisting of n singleton nodes, i.e., $E(G) = \emptyset$). Let V denote the set of all such value functions. The value of a given network G may represent any kind of total net benefit that arises when agents are connected in the specific structure of G.

The pair (N, v) is dubbed a *network game*.

For $S \subseteq N$, let G^S be the complete graph on node set S. Moreover, let $G_{|S} = \{ij \in G \mid i, j \in S\}$ be the set of edges in G connecting members of S only. In other words, $G_{|S}$ is found by starting with the graph G and then deleting edges with endnodes (agents) outside S.

Let $Z(G)$ be the set of components of G. A value function is said to be *component additive* if $v(G) = \sum_{G' \in Z(G)} v(G')$. Note that when v is component additive, there are no externalities between components in the network.

Consider a permutation $\pi \in \Pi$, and let $G^\pi = \{ij \mid i = \pi(k), \ j = \pi(l), \ kl \in G\}$. A value function is said to be *anonymous* if for any permutation $\pi \in \Pi$, we have $v^\pi(G^\pi) = v(G)$ (i.e., the value does not depend on the labeling of agents).

A network (graph) $G \in \mathcal{G}$ is said to be *efficient* if it has maximal value, that is, $v(G) \geq v(G')$ for all $G' \in \mathcal{G}$. When value is transferable between agents, this is equivalent to Pareto optimality.

To allocate the value of a given network G, $v(G)$, among the individual agents in N, we define an *allocation rule*, $\phi : \mathcal{G} \times V \to \mathbb{R}^n$, satisfying budget balance:

$$\sum_{i \in N} \phi_i(G, v) = v(G), \text{ for all } G \text{ and } v.$$

A value-allocation rule ϕ is called *anonymous* if, for any anonymous value function and for any permutation $\pi \in \Pi$, we have $\phi_{\pi(i)}(G^\pi, v^\pi) = \phi_i(G, v)$ for all $i \in N$.

Moreover, let $H(G) = \{i \in N \mid \exists j : ij \in G\}$ be the set of (active) agents with at least one connection in G. The rule ϕ is said to be *component balanced* if $\sum_{i \in H(G')} \phi_i(G, v) = v(G')$ for every G, $G' \in Z(G)$, and component-additive value function v. In other words, if a rule is component balanced, the value of any component (for which there is no connection to the other agents in N) can only be shared among agents of that component and not with any outsiders. This is only supposed to hold when v is component additive (i.e., when there is no externality between components).

The above model of a network (value) game, (N, v), is richer than the model from cooperative game theory: in a cooperative game, a single value is associated with each coalition of agents $S \subseteq N$, while in the present model many different values may be associated with the same coalition S (depending on the specific network structure by which S is connected). It is also richer than Myerson's graph-restricted game, because for instance, in the three-agent situation, Myerson's game cannot distinguish between graphs $G = \{12, 13, 23\}$ and $G' = \{12, 23\}$, in contrast to the present model.

Example 4.15 (Illustrating the concepts) Consider the three-agent case where $\mathcal{G}$ is given by the eight possible network structures illustrated below.

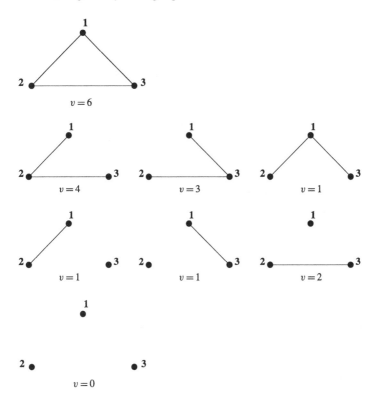

The value function v is given as shown in the figure. Compared to the cooperative game model, for instance, the (grand) coalition $\{1, 2, 3\}$ can be represented by four different network structures connecting all three agents, and consequently four different values, in the present model.

Let $G = \{12, 13, 23\}$, that is, the complete graph at the top of the figure. Projections $G_{|S}$ of G for various subsets of agents $S \subset N = \{1, 2, 3\}$ are given by $G_{|ij} = \{ij\}$ for all $i, j \in N$, $G_{|i} = \emptyset$ for all $i \in N$. For comparison, consider $G' = \{12, 23\}$, where $G'_{|13} = \emptyset$ and is otherwise identical to $G_{|S}$.

Having to specify a value for each network constellation is quite demanding (especially for practical applications). As the number of agents grows, the number of possible networks explodes: recall the complete graph G^N has $\frac{n(n-1)}{2}$ edges. Thus, the number of potential network structures is

$$\sum_{i=0}^{\frac{n(n-1)}{2}} \binom{\frac{n(n-1)}{2}}{i}. \tag{4.28}$$

That is, with four agents we have 64 different networks, with five agents we have 638 potential networks, and so on.

4.5.2 Extensions of the Shapley Value

Following the spirit of the Myerson approach (see section 1.3.1), the worth of coalition $S \subseteq N$, for a given graph G and value function v, can be represented by the value of the S-projected graph $v(G_{|S})$.[5] The Myerson value thus has a natural extension to the network model of Jackson and Wolinsky (1996), dubbed the *extended Myerson value*:

$$\phi_i^{EMV}(G, v) = \sum_{S \subseteq N \setminus \{i\}} \frac{s!(n-s-1)!}{n!} (v(G_{|S \cup \{i\}}) - v(G_{|S})) \tag{4.29}$$

for all $i \in N$. Basically the standard Shapley value is used with respect to marginal values for agent i, that is, the difference between the value of the projected network of coalitions $S \subseteq N \setminus \{i\}$ with and without the presence of agent i.

Example 4.15 (continued: Computing the extended Myerson value) To illustrate how the extended Myerson value is computed, we can reconsider example 4.15 above.

---

5. Notice that in $G_{|S}$, the complement of S, $N \setminus S$, is completely disconnected. Alternatively, the worth of coalition S could depend on how the complement is organized: see Navarro (2007) for such an approach.

For $G = \{12, 13, 23\}$, the value is $v(G) = 6$, which is divided as

$$\phi^{EMV}(G, v) = \left(\frac{10}{6}, \frac{13}{6}, \frac{13}{6} \right),$$

according to the extended Myerson value. In particular, the relevant values are $v(G_{|\{i\}}) = 0$, $i = 1, 2, 3$; $v(G_{|\{12\}}) = v(G_{|\{13\}}) = 1$; $v(G_{|\{23\}}) = 2$; and $v(G) = 6$.

Now take the graph $G' = \{12, 23\}$ with value $v(G') = 4$. Using the extended Myerson value, we get

$$\phi^{EMV}(G', v) = \left(\frac{5}{6}, \frac{11}{6}, \frac{8}{6} \right).$$

In particular, the relevant values are $v(G'_{|\{i\}}) = 0$, $i = 1, 2, 3$; $v(G'_{|\{12\}}) = 1$; $v(G'_{|\{13\}}) = 0$; $v(G'_{|\{23\}}) = 2$; and $v(G') = 4$.

Jackson and Wolinsky (1996) show that *an allocation rule ϕ satisfies component balance and equal bargaining power (a natural variation of Myerson's fairness axiom), $\phi_i(G, v) - \phi_i(G - ij, v) = \phi_j(G, v) - \phi_j(G - ij, v)$ for any ij, if and only if ϕ is the extended Myerson value $\phi^{EMV}(G, v)$, for all $G \in \mathcal{G}$, and any component-additive value function v.*

Since component balance is a rather natural fairness requirement, equal bargaining power is a quite strong axiom, despite its immediate appeal (in line with "fairness" in the Myerson model).

The following example demonstrates that the extended Myerson value has some drawbacks in the current network setting.

Example 4.16 (Drawback of the extended Myerson value)   As in Jackson (2005), consider two three-agent problems (G, v) and $(G, \bar{v})$, where $v(\{12\}) = v(\{23\}) = v(\{12, 23\}) = 1$, and $v(G) = 0$ otherwise; while $\bar{v}(G) = 1$, for all nonempty graphs. As it appears, agent 2 plays a crucial role in the problem (G, v), being the only person with whom agent 1 and 3 are able to generate any value, while agent 2 plays no particular role in the problem $(G, \bar{v})$, in which all agents are symmetric. Hence, it seems reasonable to expect that a suitable allocation rule should treat these two problems differently. However, using the extended Myerson value (e.g. with respect to the graph $\{12, 23\}$), we have $v(G_{|S}) = \bar{v}(G_{|S})$ for all $S \subseteq N$, and consequently,

$$\phi^{EMV}(\{12, 23\}, v) = \phi^{EMV}(\{12, 23\}, \bar{v}) = \left(\frac{1}{6}, \frac{2}{3}, \frac{1}{6} \right).$$

So the extended Myerson value can be completely insensitive to differences in value functions, which is obviously not a compelling feature.

Moreover, consider the following value function: $\hat{v}(\{\emptyset\}) = 0$; $\hat{v}(\{12\}) = 1 + \varepsilon$, and $\hat{v}(G) = 1$ otherwise. Basically the agents are symmetric (as $\varepsilon \to 0$), but if we look at the efficient graph $G = \{12\}$, the value of $1 + \varepsilon$ will be allocated as $(\frac{1+\varepsilon}{2}, \frac{1+\varepsilon}{2}, 0)$ using the extended Myerson value. This is arguably somewhat extreme, considering the symmetric nature of the network game (N, v).

Since agents in the Jackson-Wolinsky model are assumed to be able to form the kind of network they want, it seems reasonable to suggest that every coalition will form the network (at their discretion) that gives them the maximum value: that is, at least when v is component additive, the value associated with any coalition $S \subseteq N$ should be given by

$$w(S) = \max_{G \in G^S} v(G). \tag{4.30}$$

In particular, this means that the grand coalition N will always choose to form an efficient (value-maximizing) network.

There is a clear difference between $w(S)$ and $v(G_{|S})$, and thus also between, for instance, the Shapley value applied with respect to each of these functions. For each $i \in N$, define the Shapley value with respect to w by

$$\phi_i^{SW}(G, v) = \sum_{S \subseteq N \setminus \{i\}} \frac{s!(n - s - 1)!}{n!} (w(S \cup \{i\}) - w(S)). \tag{4.31}$$

Consider the numerical example below.

Example 4.16 (continued:) Noting that $G = \{12, 23\}$ is efficient for both v and $\bar{v}$, we can repeat the computations from example 4.16, using the new version of the Shapley value (4.31). Now we get $w(S) = \bar{w}(S)$ for all $S \neq \{13\}$, $w(\{13\}) = 0$, $\bar{w}(\{13\}) = 1$, and thus,

$$\phi^{SW}(\{12, 23\}, v) = \left(\frac{1}{6}, \frac{2}{3}, \frac{1}{6} \right)$$

and

$$\phi^{SW}(\{12, 23\}, \bar{v}) = \left(\frac{1}{3}, \frac{1}{3}, \frac{1}{3} \right).$$

So in contrast to the extended Myerson value, which resulted in the same allocation for both value functions, we now have different allocations, which arguably seems more compelling in the present case.

Moreover, in case of $\hat{v}$, we get

$$\phi^{SW}(\{12\}, \hat{v}) = \left(\frac{1}{3} + \frac{\varepsilon}{2}, \frac{1}{3} + \frac{\varepsilon}{2}, \frac{1}{3}\right),$$

which seems much more reasonable than the result of the extended Myerson value, since the allocation approaches the equal split as ε tends to 0.

Generally, for any type of network that is not necessarily efficient, Jackson (2005) suggests sticking with the coalitional values given by $w(S)$, for all $S \subseteq N$, and computing the desired solution concept, ψ, from cooperative game theory with respect to the game (N, w): then use proportional scaling afterward to obtain budget balance. That is, for any standard allocation rule ψ (the Shapley value, the nucleolus, etc.), we can define the related network game extension as

$$\phi^{\psi}(G, v) = \frac{v(G)}{w(N)} \psi(N, w). \tag{4.32}$$

Notice that $v(G)/w(N) = 1$ if G is efficient. Hence, for efficient networks, the network flexible rule ϕ^{ψ} coincides with ψ used with respect to the induced cooperative game (N, w). This has the compelling property that agents will not be in conflict over which efficient network structure to implement.[6]

In light of (4.32), it seems natural to extend to notion of the core to network (value) games, (N, v), as consisting of graphs G with feasible allocations, $y = \phi(G, v)$, in the core of the cooperative game (N, w). Formally, a network G and an allocation $y = \phi(G, v)$ are in the *core* of (N, v) if: $\sum_{i \in N} y_i \leq v(G)$, and $w(S) \leq \sum_{i \in S} y_i$ for all $S \subseteq N$. Thus, the core is nonempty if there exist a network and an allocation satisfying the above conditions.

An allocation rule ϕ is called *core consistent* if for any v such that the core is nonempty, there exists at least one graph G such that $(G, \phi(G, v))$ is in the core. It is easy to demonstrate that extensions building on the Shapley value fail core consistency, and Jackson (2005) therefore suggests looking at a version of the nucleolus (dubbed "networkolus").

We shall return to a noncooperative version of network (formation) games in section 5.3.

---

6. Navarro (2010) suggests some variations of this type of network flexible rule based on identified externalities.

4.6 Flow Problems

Networks are often distribution systems in which some homogeneous resource (e.g, water, electricity, gas) is transferred in a directed flow from supply to demand nodes via different locations involving different subnetwork owners. We will therefore turn to network flows and the kind of allocation problems that arise in this context. In particular, we will study loss allocation in energy transmission networks. This problem poses a classic challenge to regulation and can potentially have major impacts on system efficiency and development.

But flow models have also been a popular topic in the operations research literature, which offers several models and methods to determine optimal flows. However, allocation problems have not been the prime focus of that literature. In fact, Meggido (1974) presents one of the first studies of fair allocation in flow models: a planner has to distribute amounts of a homogeneous resource among agents located at multiple sources, and these agents are going to ship their allocations, through a network with capacity constraints, to other agents located at multiple sinks. The question is how a planner should distribute the flow fairly among the agents. When equal division is infeasible due to network constraints, Meggido (1974) suggests opting for flows that maximize the minimum amount delivered from individual sources or supplied to individual sinks.

However, the best-known model is the max-flow model of Kalai and Zemel (1982a, b): a flow of value is shipped from a single source to a single sink, given the capacity constraints of each arc in the network. Agents control (own) individual arcs and have to cooperate in order to obtain efficiency in the form of a max-flow, given capacity constraints. The value of the max-flow must subsequently be allocated among the agents based on the arc ownership structure.[7]

4.6.1 Transmission Networks

Following Bergantiños et al. (2017), we start out by considering a network flow problem based on energy losses in gas transmission networks.

Transmission networks are typically owned by many different agents, called *haulers* in case of gas transmission. A central authority is managing the network, so haulers have basically no influence on the flow or the structure of the network itself, which therefore can be considered as fixed.

---

7. For of max-flows, we have no interest in generalizing to multiple sources and sinks, since we can always add a "supersource" and a "supersink" along with arcs leading from these to a set of (original) multiple sources and sinks.

Transmission networks consist of multiple supply nodes (sources) and demand nodes (sinks), as well as intermediate nodes where flows intersect. When gas is transported from supply nodes to demand nodes, there is a loss of gas in the system, which is typically around 0.2 percent in high-pressure networks.[8]

Due to concerns about system reliability, the expected gas loss will be withheld at entry points in the network, which in turn implies an added cost for the responsible haulers. Since the loss is a common loss for the entire network system, it is reasonable to allocate it among all the haulers. The managing network authority decides how the loss is allocated among haulers. Using some allocation rule, each hauler h is assigned ex ante an allowed loss, say, A_h. The actual (ex post) loss for hauler h is determined by the gas balance between entry and exit points in the subnetwork owned by hauler h, say, L_h.

In a European context, allocation rules and payment mechanisms are subject to EU regulation. For instance, in accordance with the EU regulation, the Spanish regulatory authority uses a mechanism whereby haulers are penalized for losing more than what is allowed by the central allocation (i.e., if $L_h - A_h > 0$). In this case, a hauler has to pay $p(L_h - A_h)$, where p is the unit price of gas. But if a hauler is losing less than the allowed amount (i.e., if $L_h - A_h < 0$), the hauler will receive payment $\frac{p}{2}(A_h - L_h)$. Since haulers have no (or very little) influence on network operation, their only incentive is to reduce the loss in their own part of the system, which is aligned with the regulator's interest in increased system efficiency.

From the structure of the payment mechanism, it is clear that the allocation of allowed losses, A_i, matters for the haulers: haulers prefer as much allowed loss as possible. Regulation therefore also states that the applied allocation rule has to satisfy a number of normative fairness properties, such as nondiscrimination and cost reflectivity. Moreover, the allocation rule should induce competition among haulers. Consider the following example, which illustrates some of the challenges in finding a compelling allocation.

Example 4.17 (Allocating loss penalties)   Assume for simplicity that only the size of the flow matters for the transmission loss in haulers' arcs. Consider the network below with two supply nodes (s_1, s_2), two demand nodes (t_1, t_2), and one intermediate node (z). Suppose three haulers, $H = \{h_1, h_2, h_3\}$, own the arcs in the system. Hauler 1 owns arcs 1 and 2, with flow $f_1 = 200$ and

8. Bergantiños et al. (2017) report that in the Spanish gas transmission network, a gas loss of 0.2 percent corresponds to a monetary loss of approximately 25 million euros in the system as a whole.

$f_2 = 600$, respectively. Hauler 2 owns arc 3, with flow $f_3 = 800$; and hauler 3 owns arc 4, with flow $f_4 = 400$.

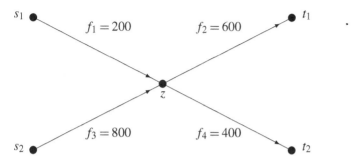

In total, 1,000 units of gas are transported in the network with an estimated loss of $(0.002 \times 1000=)$ 2 units. Notice that if we allow each hauler to lose 0.2 percent of the gas at each entry point in the hauler's system, hauler 1 is allowed to lose 1.6 units $(=0.002 \times 200 + 0.002 \times 600)$, hauler 2 is allowed to lose 1.6 units $(=0.002 \times 800)$, and hauler 3 is allowed to lose 0.8 units $(=0.002 \times 400)$. In total this becomes 4 units, which exceeds the 2-unit loss in the system as a whole. A straightforward loss computation for each hauler is therefore not an option, since it will not produce budget balance.

Ideally we can assign a flow to each hauler for which the hauler is held liable. Then the system loss can, for instance, be allocated in proportion to such hauler-specific flows. But as it appears from the graph above, this is not straightforward, since there is no uniquely defined way to do so. One possible proxy, which is easy to apply in practice, is the total amount of gas entering a hauler's subnetwork (i.e., the total net amount of gas leaving all the nodes in the hauler's subnetwork): in the above network this is $200 + (600 - 200) = 600$ units for hauler 1, 800 units for hauler 2, and 400 units for hauler 3. Proportional allocation therefore gives $(0.66, 0.88, 0.44)$.

Another approach traces the amount of gas going from each supply node to each demand node. If such a tracing can be done, we can decompose the problem into several single-flow problems that are much easier to handle, since haulers who own arcs along the single flow can then share the loss equally, in proportion to length of their pipes, or some similar measure. But the complexity created by the intermediate nodes prevents exact tracing, and we therefore have to rely on mechanical methods. One such method allocates the outflow of an intermediate node in proportion to the incoming flows. In the above graph, we see that 20 percent of the inflow to node z comes via arc 1, and 80 percent via arc 3; thus, the outflow of 600 units in arc 2 can be decomposed proportionally as $0.2 \times 600 = 120$ units from arc 1, and $0.8 \times 600 = 480$ units

from arc 3. Likewise the outflow of 400 units in arc 4 can be decomposed proportionally as $0.2 \times 400 = 80$ units from arc 1, and $0.8 \times 400 = 320$ units from
arc 3. The entire flow can be decomposed into four single flows (two for each
incoming arc):

- a flow of 120 units via the path $s_1 \to z \to t_1$,
- a flow of 80 units via the path $s_1 \to z \to t_2$,
- a flow of 480 units via the path $s_2 \to z \to t_1$, and
- a flow of 320 units via the path $s_2 \to z \to t_2$.

In each single flow, the loss is 0.2 percent of flow size, so the losses in
the four flows are $(0.24, 0.16, 0.96, 0.64)$, respectively. Suppose that for each
flow, we allocate that loss equally among haulers having arcs involved. Thus,
we obtain the final loss allocation given by $(0.8, 0.8, 0.4)$: this time favoring
hauler 1, who is allowed to lose 0.8 units versus 0.66 units before.

4.6.1.1   The model

To model the loss-allocation problem, we let H denote the set of haulers and
L the total system loss that has to be allocated. Moreover, we assume that
each arc is characterized by both a flow f and a volume v. In practice it is
complicated to say exactly how flow and volume affect the size of the loss, but
loosely speaking, an increasing flow will increase loss, and increasing volume,
through its correlation with the length of pipes, tends to increase loss as well.

A *gas-loss problem* is a 5-tuple $\gamma = (G, f, v, \mathcal{H}, \alpha)$, where $G = (N, E)$ is
a graph representing the transmission network (without directed cycles), $f = \{f_e\}_{e \in E}$ is a profile of flows in every arc, $v = \{v_e\}_{e \in E}$ is a profile of volumes in
every arc, $\mathcal{H} = \{E_h\}_{h \in H}$ is a hauler structure partitioning the set of arcs E into
subsets owned by different haulers (that is, $\cup_{h \in H} E_h = E$), and $\alpha \in [0, 1]$ is the
loss percentage in the system. Denote by Γ the set of gas-loss problems.

An allocation rule $\phi : \Gamma \to \mathbb{R}^H$ is a function assigning a loss to every hauler
under budget balance: $\sum_{i \in H} \phi_i(\gamma) = L$.

For every node $j \in N$, the difference $\sum_{(j,l) \in E} f_{(j,l)} - \sum_{(l,j) \in E} f_{(l,j)}$ is the
netflow leaving node j (i.e., the difference between the total flow leaving
node j and the total flow arriving at node j). Recall from chapter 1, that
the net flow is 0 if j is an intermediate node, positive if j is a supply node,
and negative if j is a demand node.

Denote by

$$F_h = \sum_{j \in N} \max \left\{ \sum_{(j,l) \in E_h} f_{(j,l)} - \sum_{(l,j) \in E_h} f_{(l,j)}, 0 \right\} \qquad (4.33)$$

the total amount of gas entering hauler h's subnetwork. Let $S \subset N$ be the set of supply nodes, and $T \subset N$ the set of demand nodes in G. For every pair $(s, t) \in S \times T$, let $P(s, t)$ be the set of paths from s to t in G, and let $P(S, T) = \cup_{(s,t) \in S \times T} P(s, t)$ be the set of all paths from supply to demand nodes. Moreover, for each path $p \in P(S, T)$, let f^p be the flow assigned to p using proportional tracing (i.e., by decomposing every outflow from a given node in proportion to the size of inflows, as illustrated in example 4.17 above).

4.6.1.2 Some rules
We can now define some allocation rules that are discussed in the literature. The first is called the *flow rule*: it allocates the total loss in proportion to the total amount of gas entering hauler h's subnetwork,

$$\phi_h^{FR}(\gamma) = \frac{F_h}{\sum_{i \in H} F_i} L \tag{4.34}$$

for all $h \in H$. Bergantiños et al. (2017) report that the flow rule has been used by the Spanish regulatory authority since 2014.

Using the proportional tracing method, transmission can be decomposed into single flows as given by $P(S, T)$. For every such single flow p, we can now allocate its estimated loss, αf^p, among the arcs in p in proportion to arc volume. In this way, the *proportional tracing rule* assigns loss shares given by

$$\phi_h^{PTR}(\gamma) = \alpha \sum_{e \in E_h} \sum_{p \in P(S,T):e \in p} f^p \frac{v_e}{\sum_{e \in p} v_e} \tag{4.35}$$

to every hauler $h \in H$. Notice that in example 4.17 above, we used a tracing rule where path flow is shared equally among haulers owning arcs in the flow, while version (4.35) allocates in proportion to arc volume.

But the loss can also be allocated arc-by-arc, using, for instance, the *arc rule*. In this case the total loss is shared in proportion to the product of an arc's flow an volume as

$$\phi_h^{AR}(\gamma) = \frac{\sum_{e \in E_h} f_e v_e}{\sum_{e \in E} f_e v_e} L \tag{4.36}$$

for every hauler $h \in H$.

Example 4.17 (continued: Computing loss shares) Recall the problem in example 4.17 above. Now, suppose that the volumes of arcs 1–4 are given by $v_1 = 100$, $v_2 = 50$, $v_3 = 150$, and $v_4 = 100$. Since the flow rule, defined in (4.34), does not depend on volume, we have already computed the loss allocation, given by

$$\phi^{FR} = (0.66, 0.88, 0.44).$$

Using the proportion tracing rule, defined in (4.35), we have to reconsider the four possible single flows. The path $s_1 \to z \to t_1$ is fully owned by hauler 1, who therefore carries its loss of 0.24. For the path $s_1 \to z \to t_2$, the first arc is owned by hauler 1, while the last arc is owned by hauler 3: since the volume is 100 for both arcs, haulers 1 and 3 share the flow loss of 0.16 equally. For the path $s_2 \to z \to t_1$, the first arc is owned by hauler 2 and the last arc by hauler 1: since the volumes are 150 and 100, respectively, the flow loss of 0.96 is shared in ratios $(0.75, 0.25)$. Finally, the path $s_2 \to z \to t_2$ is shared between haulers 2 and 3: since the respective volumes are 150 and 100, the flow loss of 0.64 is shared in ratios $(0.6, 0.4)$. Thus, we have

$$\phi^{PTR} = (0.56, 1.10, 0.34).$$

Compared to the situation where volumes are disregarded and losses split equally among haulers owning arcs in the respective single flows, hauler 2 is now allowed to lose more. This is because the volume of hauler 2's sub-network (arc 3) is the highest ($v_3 = 150$), indicating that hauler 2's network is longer, and therefore hauler 2 should be allowed to lose more than the other haulers.

Finally, using the arc rule, defined in (4.36), we have

$$\phi^{AR} = (0.48, 1.14, 0.38).$$

Again the fact that the volume of hauler 2's arc is the highest allows hauler 2 to lose much more than either hauler 1 or hauler 3.

Bergantiños et al. (2017) analyze a series of normative requirements related to cost reflectivity, nondiscrimination, and competition. We will briefly consider a few of these below.

Concerning cost reflectivity, it seems reasonable to require that if two problems only differ on arcs without flow, the loss allocation should be the same in both cases.

Independence of unused arcs Consider two problems, γ and $\bar{\gamma}$, where $H = \bar{H}$, and for each $h \in H$, $\bar{E}_h = E_h \setminus \hat{E}$, where $\hat{E} \subset E$ satisfies the conditions that for each $e \in E \setminus \hat{E}$, $\bar{f}_e = f_e$ and $\bar{v}_e = v_e$ and for each $e \in \hat{E}$, $f_e = 0$. Then $\phi(\gamma) = \phi(\bar{\gamma})$.

This property is satisfied by all three allocation rules defined above. Moreover, it also seems reasonable to require that the loss allocated to a given hauler should be independent of who owns the arcs outside the hauler's subnetwork.

Ownership independence Let two problems γ and $\bar{\gamma}$ differ only with respect to the hauler structure $\mathcal{H}$ and $\overline{\mathcal{H}}$, such that for $h_1, h_2 \in H$, and $e \in E$, we have $\bar{E}_{h_1} = E_{h_1} \setminus \{e\}$ and $\bar{E}_{h_2} = E_{h_2} \cup \{e\}$, and $\bar{E}_h = E_h$ otherwise. Then, for each $h \in H \setminus \{h_1, h_2\}$, $\phi_h(\gamma) = \phi_h(\bar{\gamma})$.

Ownership independence is satisfied by the proportional tracing rule and the arc rule, but *not* by the flow rule used by the Spanish regulator (indeed, in example 4.17 above, suppose hauler 1 sells arc 1 to hauler 2. The new profile of hauler flows becomes $F = (600, 1000, 400)$ so the loss allocated to hauler 3 now becomes $\phi_3^{FR} = \frac{400}{2000}2 = 0.4 < 0.44$).

Concerning nondiscriminatory properties, different versions of symmetry seem relevant. For instance, if two haulers are symmetric in terms of their arcs having the same flow and volume, they should be allowed the same loss.

Symmetry on arcs Let γ and $h_1, h_2 \in H$ be such that $E_{h_1} = \{e\}$, $E_{h_2} = \{\bar{e}\}$, $f_e = f_{\bar{e}}$, and $v_e = v_{\bar{e}}$. Then $\phi_{h_1}(\gamma) = \phi_{h_2}(\gamma)$.

Symmetry on arcs is satisfied by both the flow and arc rules, but *not* by the proportional tracing rule. (Indeed, let $H = \{h_1, h_2, h_3\}$, $S = \{s\}$, $T = \{t_1, t_2\}$, and $N \setminus (S \cup T) = \{z_1, z_2\}$. Moreover, four arcs have flows $f_{sz_1} = f_{z_1t_1} = f_{sz_2} = f_{z_2t_2} = 500$ and volumes $v_{sz_1} = v_{z_1t_1} = v_{z_2t_2} = 100$, $v_{sz_2} = 50$. Suppose that h_1 owns arcs sz_1 and sz_2, h_2 owns arc z_1t_1, and h_3 owns arc z_2t_2. Arcs z_1t_1 and z_2t_2 are symmetric, and yet $\phi_{h_2} = 0.5 \neq 0.67 = \phi_{h_3}$.)

Regarding competition, the regulators' choice of allocation rule must ensure that it is disadvantageous for haulers to merge (and thereby reduce competition). This implies that the allocated loss to a merged hauler (which can be seen as the loss it is allowed to have) should be smaller than the sum of allocated losses to each of the merging haulers: all three allocation rules above ensure that it is indeed smaller. In particular, notice that ownership independence implies this kind of merger disadvantage.

Although all three rules have their pros and cons, Bergantiños et al. (2017) end up rating both the proportional tracing rule and the arc rule superior to the currently used flow rule. They further provide axiomatic characterizations of these two rules, building on some of the properties considered above.

4.6.2 Max-Flow Problems

As mentioned in chapter 1, the max-flow problem is a classic in the operations research literature, pioneered by the work of Ford and Fulkerson (1962). Since then, several algorithms have appeared for finding a max-flow in a given graph with given capacity constraints (see, e.g., Kozen, 1992). The allocation issue arises when agents control individual arcs and need to coordinate efforts in order to obtain a (common) max-flow of, say, revenue, which is then subsequently distributed among the agents.

4.6.2.1 The model

Recall from chapter 1 that a *max-flow problem* consists of a digraph $G = (N, E)$, where N is a set of nodes and E is a set of arcs, with two distinguished nodes: a source, $s \in N$, and a sink, $t \in N$. Each arc $e \in E$ has a maximum (finite) capacity $c_e \geq 0$ and is controlled by a given agent among a finite set M of agents: some agents may control (own) several arcs. The planner wants to direct as much flow as possible from the source to the sink subject to the capacity constraints. Denote by $F(M)$ the maximum source-to-sink flow in case of agent set M. There may be several different ways to obtain a max-flow for a given problem.

Max-flow problems can understood as models of integrated production, communication, or transportation systems, where the flow generates revenue. For instance, a problem may consist of routing as many packets as possible through the network. The planner, representing the agents as a group, wants to maximize revenue, but this requires cooperation of the individual agents controlling the arcs. In other words, we are back to allocation problems. Because we are dealing with revenue sharing, it is also (implicitly) assumed that agents want as much flow through their arcs as possible.

4.6.2.2 The max-flow game

Since agents have to coordinate their efforts to obtain a max-flow, it seems natural to apply the tools of cooperative game theory, as originally suggested in Kalai and Zemel (1982a, b). For any coalition of agents $S \subseteq M$, we can define an S-restricted max-flow problem by letting G^S be the digraph obtained from G by keeping all the nodes N but only using arcs E^S controlled by members of S. Let $F(S)$ be the maximal source-to-sink flow in the S-restricted problem. Thus, the pair (M, v), where

$$v(S) = F(S) \quad \text{for all} \quad S \subseteq M, \tag{4.37}$$

constitutes a *max-flow game*. We can think of $v(S)$ as the maximum potential revenue obtainable by coalition S without the cooperation of agents in $N \setminus S$.

Example 4.18 (A max-flow problem and its associated max-flow game)
Three agents, Ann, Bob, and Carl, control arcs A, B, and C, respectively, in the digraph G given by the figure below, where numbers indicate the maximum capacity of the arc.

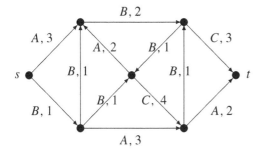

It is easy to see that the max-flow becomes 3 (recall example 1.6 in chapter 1, where the Ford-Fulkerson algorithm was used to find the max-flow). So in total Ann, Bob, and Carl are able to ship 3 units from the source s to the sink t. Clearly, the max-flow of any individual agent is 0, since nobody has full control of a path from source to sink. It is also clear that the coalition {Ann, Carl} does not control enough arcs to form a path from source to sink and consequently has a max-flow of 0. The coalition {Ann, Bob} has a max-flow of 1 (using the lower path), and the coalition {Bob, Carl} also has a max-flow of 1, for example, using the path shown below.

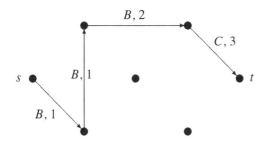

The associated max-flow game becomes

$$v(A) = v(B) = v(C) = 0,$$

$$v(A, B) = v(B, C) = 1, v(A, C) = 0,$$

$$v(A, B, C) = 3.$$

Clearly, the core is nonempty and consists of allocations (x_A, x_B, x_C) satisfying $x_A + x_B + x_C = 3$, and $0 \leq x_A \leq 2, 0 \leq x_B \leq 3, 0 \leq x_C \leq 2$.

Given a max-flow problem, a *cut* is a set of arcs $E^* \subseteq E$ such that the digraph $G - E^*$ has a max-flow of 0. The capacity of a cut is the sum of the capacities of the arcs in the cut: $\sum_{e \in E^*} c_e$. Recall from chapter 1 that the capacity of any cut is greater than or equal to the max-flow, and there exists a minimal cut (i.e., a cut with capacity equal to the max-flow).

Consider a given max-flow problem, and let E^* be a minimal cut. For each agent $i \in M$, let c^i be the total capacity of arcs in E^* controlled by agent i. By the Ford-Fulkerson result mentioned above,

$$F(M) = v(M) = \sum_{i \in M} c^i.$$

Thus, $c = (c^1, \ldots, c^m)$ is actually an allocation in the associated max-flow game (M, v).

Moreover, letting $E^*(S)$ be all the arcs in the minimal cut E^* controlled by agents in $S \subseteq M$, it is easy to see that $E^*(S)$ is a cut in the S-restricted max-flow problem (indeed, every path going from s to t using only arcs controlled by members in S must contain an arc from $E^*(S)$). Thus, using the Ford-Fulkerson result, we have

$$F(S) = v(S) \le \sum_{e \in E^*(S)} c_e = \sum_{i \in S} c^i.$$

In other words, the allocation $c = (c^1, \ldots, c^m)$ satisfies the standalone core conditions of the derived max-flow game (M, v). Thus, this game must be balanced. In fact, every max-flow game is totally balanced, since it is easy to show that every subgame of a max-flow game is itself a max-flow game, and hence it follows that every subgame has a nonempty core.

As shown in Kalai and Zemel (1982a), the inverse result also holds true: a game is totally balanced if and only if it is a max-flow game.

Example 4.18 (continued: A minimal cut allocation is a core allocation) In the network of example 4.18 above there exists a unique minimal cut as shown in the figure below (using the symbol $//$)

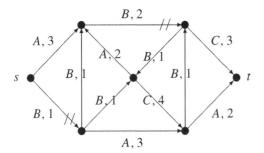

Note that both arcs in the minimal cut are controlled by Bob, so using a minimal cut allocation, we have $x = (0, 3, 0)$, which is an extreme core allocation (allocating the maximum amount to Bob and the minimum amount to Ann and Carl, given the core constraints). This allocation is hardly fair, since Bob cannot obtain any flow on his own: he has to cooperate with either Ann or Carl (or both). Moreover, Bob is always the one restricting the flow, because his arc has (weakly) lower capacity than both Ann's and Carl's.

For comparison, the Shapley value is given by the allocation $x^{Sh} = \left(\frac{5}{6}, \frac{4}{3}, \frac{5}{6}\right)$, which is a core allocation, and even the equal spilt $(1, 1, 1)$ is a core allocation in this case.

As demonstrated by the example above, a minimal cut allocation may be quite unfair, somewhat reminiscent of a Bird allocation in the MCST problem. Also, like Bird allocations, there may be many minimal cut allocations for a given max-flow problem. Thus, we can either define a set-valued solution concept or a standard allocation rule by taking, for instance, a (weighted) average of all the minimal cut allocations for a given problem. Reijnierse et al. (1996) analyze the set-valued minimal cut solution in the special case of *simple flow games*, where the set of agents is identified by the set of arcs, and the maximal capacity of every arc is normalized to 1. On the class of simple flow games, Reijnierse et al. characterize the set-valued minimal cut solution by versions of *consistency* and *converse consistency*.

The minimal cut allocations can be quite unfair, but even the core itself may conflict with fairness, as illustrated by the following example.

Example 4.19 (All allocations in the core seem unfair) Consider the problem with four agents $\{A, B, C, D\}$ illustrated below, where the max-flow is 2 and there is a unique minimal cut ($//$).

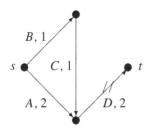

The associated max-flow game is given by

$$v(i) = 0 \text{ for } i \in \{A, B, C, D\},$$
$$v(AD) = 2, \quad v(ij) = 0 \text{ otherwise,}$$

$$v(ABC) = 0, \quad v(ABD) = v(ACD) = 2, \quad v(BCD) = 1,$$

$$v(ABCD) = 2.$$

This game is totally balanced but is not convex: $v(ACD) + v(BCD) = 3 > 2 = v(CD) + v(ABCD)$. The core is given by allocations

$$\alpha(1, 0, 0, 1) + (1 - \alpha)(0, 0, 0, 2), \quad \alpha \in [0, 1].$$

Thus, core allocations guarantee 1 unit for agent D and let A and D share the remaining 1 unit, with nothing for agents B and C. Using the minimal cut allocation, we get the extreme core allocation $(0, 0, 0, 2)$. But giving nothing to agents B and C is not compelling, since one way to reach the max-flow of 2 would be to transfer 1 unit via B and C, and 1 unit via A, adding up to 2 units via D's arc. Indeed, the Shapley value of the game (which is not a core allocation in this case) is given by

$$x^{Sh}(v) = \left(\frac{3}{4}, \frac{1}{12}, \frac{1}{12}, \frac{13}{12} \right),$$

acknowledging that agents B and C ought to share the benefits, albeit by a significantly smaller amount than for A and D.

Moreover, note that if we raise the capacity of the arcs controlled by B and C from 1 to 2 units, as illustrated below,

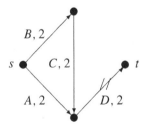

we get a new max-flow game $\bar{v}$, where $\bar{v}(BCD) = 2$, and $\bar{v}(S) = v(S)$ otherwise. This time the core consists of a single allocation, $(0, 0, 0, 2)$, giving everything to agent D, even though going from v to $\bar{v}$ increases the "power" of agents B and C: the core reacts by an extreme punishment to A, but not by adding to the payoffs of B and C. Again we can compare with the more compelling Shapley value, which result in the allocation

$$x^{Sh}(\bar{v}) = \left(\frac{1}{2}, \frac{1}{6}, \frac{1}{6}, \frac{7}{6} \right),$$

where A's payoff is decreased from 0.75 to 0.5, while the payoffs of the remaining agents have increased.

In general when the set of agents is identical to the set of arcs E (i.e., each agent controls one and only one arc, as in this example), an arc (agent) $e \in E$ is called redundant if $v(E \setminus \{e\}) = v(E)$. As observed in Sun and Fang (2007) an arc/agent is redundant if and only if its payoff is 0 in all core allocations (i.e., $x_e = 0$ for all $x \in core(E, v)$). In the graph above, agents A, B, and C are all redundant (and hence get 0 payoff in the core), yet cooperation with either A or coalition $\{B, C\}$ is necessary to obtain the max-flow of 2, so the core seems less compelling in this case. This example further demonstrates that any allocation rule that shares the max-flow among nonredundant agents is not compelling.

The next example demonstrates some further weaknesses of the game-theoretic approach.

Example 4.20 (Further weaknesses of the game-theoretic approach) Consider the following max-flow problem G with two agents, Ann and Bob.

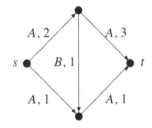

Clearly, Bob's arc can never be part of a max-flow, and the associated max-flow game becomes

$$v(A) = 3, \quad v(B) = 0, \quad v(AB) = 3.$$

Now change the direction of Bob's arc, so we get the graph G' below.

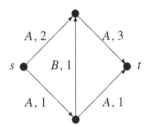

Bob's arc can now be part of a max-flow, but the associated max-flow game remains unchanged. Thus, no game-theoretic solution concept can guarantee positive payment to agents with arcs that could potentially be part of a max-flow.

Furthermore, game-theoretic solution concepts (here the Shapley value) may reward agents for having completely useless arcs in terms of the max-flow, as illustrated in the graph below.

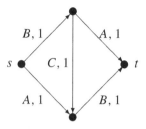

There are three agents, Ann, Bob, and Carl, and Carl's arc is clearly useless, since it will never be part of a max-flow. Yet because it may be used in a suboptimal flow together with Bob's arcs, the associated max-flow game becomes

$$v(A) = v(B) = v(C) = 0, \; v(AB) = 2, v(AC) = 0, v(BC) = 1,$$

$$v(ABC) = 2,$$

with Shapley value $\left(\frac{2}{3}, \frac{7}{6}, \frac{1}{6}\right)$. So even though Carl's arc can only prevent obtaining a max-flow, he is rewarded by the Shapley value, because he has one positive marginal value $v(BC) - v(B) = 1$. A much more compelling solution would be for Ann and Bob to share equally $\left(\frac{1}{2}, \frac{1}{2}, 0\right)$ in this case, corresponding to the unique Lorenz-maximizing element of the core.

An alternative to the game-theoretic approach could be to decompose given max-flows into single flows by using tracing methods, as described in section 4.6.1. Based on these methods, various kinds of allocation rules can be defined along the lines suggested in section 4.6.1. For instance, it seems natural to suggest that agents controlling arcs in a given single flow should share the value of that flow equally or in proportion to their arc capacity. Another line of approach defines a liability index for each edge—for instance, the flow that necessarily has to pass through the edge in any max-flow, the maximum flow that can pass through the edge in any max flow, or some convex combination of those. Then the max flow can be shared according to such liability indices, for example, in proportion. Such approaches have not been pursued in the literature so far.

4.7 Exercises

4.1. Consider the digraph G below. Is it (directed) cycle free?

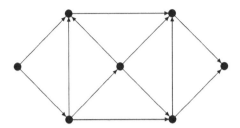

Interpret G as a hierarchy. Use a fixed-fraction transfer rule ζ (4.1) to transfer revenue in the associated revenue-sharing problem (say, revenue is normalized to 1 for each agent). Interpret the revenue transfer as a flow. What is the role of source and sink?

4.2. Consider the graph below with four nodes $\{1, 2, 3, 4\}$ and three edges. Each edge carries a separate cost.

Three agents have the following connection demands: Ann wants to connect nodes 1 and 2; Bob wants to connect nodes 1 and 3; and Carl, nodes 1 and 4. Is there any redundancy in this problem? Find $\overline{\mathcal{D}}^i$ for all agents. Compute the liability indices θ^0, θ^1, and θ^p for all agents and all edges. Compare the way edge costs are shared using proportional ratios with respect to the three liability indices. Moreover, compare these results with the result of using various game-theoretic solutions to the airport game (see chapter 2).

Next, consider four nodes $\{1, 2, 3, 4\}$ and the complete six-edge graph illustrated below. Ann wants to connect nodes 1 and 2. Bob wants to connect nodes 3 and 4.

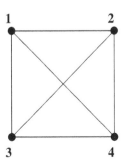

Find Ann's and Bob's minimal subgraphs, and compute liability indices θ^1 and θ^p for all edges. Comment on the result of sharing costs in proportion to these indices.

Now assume that Ann wants 2-connectivity (i.e., wants to connect nodes 1 and 2 by two edge-disjoint paths). Find her minimal subgraphs in this case, and compare her liability indices θ^1 and θ^p with the 1-connectivity situation above.

4.3. Prove that cost ratios (shares) of the proportional form (4.5) based on the probabilistic liability index (4.9) satisfy all axioms of theorem 4.3 except for irrelevance of supplementary edges. Provide an example showing that irrelevance of supplementary edges is violated.

4.4. Consider, as in section 4.2.5, cost-allocation problems with limited reliability $(M, G, \{\mathcal{D}^j\}_{j \in M}, p, c)$, and let p be IID with a probability of success q for every edge. Compute and compare cost allocations using the ex post rule, the expected counting liability rule, and the needs priority rule in the following two situations with three agents $M = \{1, 2, 3\}$ and three edges $E = \{a, b, c\}$:

Substitutable needs: $\overline{\mathcal{D}}^1 = \{a\}, \overline{\mathcal{D}}^2 = \{a, b\}, \overline{\mathcal{D}}^3 = \{a, b, c\}$;

Complementary needs: $\overline{\mathcal{D}}^1 = \{a\}, \overline{\mathcal{D}}^2 = \{ab\}, \overline{\mathcal{D}}^3 = \{abc\}$.

4.5. For allocation problems with limited reliability, $(M, G, \{\mathcal{D}^j\}_{j \in M}, p, c)$, define an associated cooperative game with expected coalitional values. Does the Shapley value of this game satisfy cost additivity and/or independence of timing?

4.6. Show that the serial responsibility rule (4.21) violates consistency. Does the serial responsibility rule coincide with the Shapley value of the associated cooperative game?

4.7. (Henriet and Moulin, 1996) n agents want connection to a source (e.g., a central switching machine) in a communication network. There is a fixed cost, given by the profile $c = (c_1, \ldots, c_n)$, of connecting the agents to the network (local wires, switching cards, etc.). Let x_{ij} be the amount of traffic between agents i and j (number of hours of communication between i and j per month). Let X denote the $n \times n$ traffic matrix with $x_{ij} = x_{ji}$ and $x_{ii} = 0$. Let the pair (c, X) denote the allocation problem.

Discuss the interpretation of the following two allocation rules:

$$\phi_i^{PC}(c, X) = c_i \qquad\qquad (4.38)$$

for all $i = 1, \ldots, n$, and

$$\phi_i^{EC}(c, X) = \sum_j \frac{x_{ij}}{\sum_i x_{ij}} c_j \tag{4.39}$$

for all $i = 1, \ldots, n$.

Do they satisfy the standard properties of cost additivity and scale invariance?

Now consider the following two properties and provide an interpretation of them.

Sustainability Consider two traffic matrices, X and X', where $x_{ij} \neq x'_{ij}$, and $x_{nm} = x'_{nm}$ otherwise. Then for all coalitions $S \supseteq \{i, j\}$,

$$\sum_{l \in S} \phi_l(c, X) - \sum_{l \in S} \phi_l(c, X') \leq c_i + c_j.$$

No transit Consider a given traffic matrix X, and let X' be constructed from X by transferring an amount of traffic $\gamma \geq 0$ from $\{i, k\}$ via j, that is, $x'_{ik} = x_{ik} - \gamma$, $x'_{ij} = x_{ij} + \gamma$, and $x'_{jk} = x_{jk} + \gamma$. Then for all $S \supseteq \{i, j\}$ and $k \notin S$,

$$\sum_{l \in S} \phi_l(c, X) - \sum_{l \in S} \phi_l(c, X') \leq c_i + c_j.$$

Show that both allocation rules above satisfy sustainability and no transit. Find a natural allocation rule that satisfies neither.

Finally, show that (for $n \geq 4$) a cost-allocation rule ϕ satisfies sustainability, no transit, cost additivity, and scale covariance if and only if, for each $j \in N$ there is a number, $\lambda_j \in [0, 1]$, such that

$$\phi_i(c, X) = (1 - \lambda_i)c_i + \sum_{k \neq i} \lambda_k \frac{x_{ik}}{\sum_i x_{ik}} c_k,$$

for all $i \in N$. [Hint: It is well known that cost additivity implies linearity of the allocation rule, i.e. $\phi_i(c, X) = \sum_{j \in N} \alpha_j^i(X) c_j$, where $\alpha_j^i \geq 0$ and $\sum_{i \in N} \alpha_j^i(X) = 1$. Show how sustainability and no transit can be used to give further structure to the weights α_j^i.]

4.8. (Bogomolnaia et al., 2010) Consider capacity networks and the game defined by (4.24). Prove that this game is concave. An allocation rule is monotonic if payments are weakly increasing in the capacities k_{ij} for all $i, j \in N$.

Show that the Shapley value of the game based on the irreducible matrix satisfies monotonicity.

4.9. (Jackson and Wolinsky, 1996) In chapter 5 we shall return to the network games model of Jackson and Wolinsky. In particular, we will consider the notion of *pairwise stability:* A graph G is pairwise stable with respect to v and ϕ if

1. for all $ij \in G$, $\phi_i(G, v) \geq \phi_i(G - ij, v)$ and $\phi_j(G, v) \geq \phi_j(G - ij, v)$; and

2. for all $ij \notin G$, if $\phi_i(G, v) < \phi_i(G + ij, v)$, then $\phi_j(G, v) > \phi_j(G + ij, v)$.

The graph G is said to *defeat* G' if $G' = G - ij$ and condition 1 is violated for ij, or if $G' = G + ij$ and condition 2 is violated for ij. Consider, and comment on, the following axioms:

Pairwise monotonicity ϕ is pairwise monotonic if G' defeats G implies that $v(G') > v(G)$.

Independence of potential links ϕ is independent of potential links if $\phi(G, v) = \phi(G, w)$ for all G and v, w such that there exists $j \neq i$ so that v and w agree on every graph except $G + ij$.

Prove that the equal-split rule (assigning the payment $v(G)/n$ to every agent) is uniquely characterized by anonymity, pairwise monotonicity, and independence of potential links. Comment on the connection to theorem 4.2.

4.10. (Kalai and Zemel, 1982b) A *simple* max-flow problem is a max-flow problem where the set of arcs is identified by the set of agents (i.e., $|E| = |N|$, and each arc is controlled by a different agent), and the capacity of each arc is normalized to 1. Consider the associated simple max-flow game and prove that the extreme points of the core are precisely the points $x = \{x_i\}_{i \in N}$ such that

$$x_i = \begin{cases} 1 & \text{if } i \in E^*, \\ 0 & \text{otherwise,} \end{cases}$$

where E^* is a minimal cut.

4.11. (Reijnierse et al., 1996) An arc in a max-flow problem is called *public* if no agent controls it. Show that if public arcs are allowed, the minimal cut allocation is still a core member for all minimal cuts that do not contain public arcs. Moreover, provide an example of a graph where no such minimal cuts exist (i.e., a graph where all minimal cuts contain at least one public arc).

4.12. (Rhys, 1970) Consider a case where four (freight handling) terminals T_i, $i = 1, 2, 3, 4$, are used by four services S_i, $i = 1, 2, 3, 4$. The service S_1

uses (connects) terminals T_1 and T_2; S_2 uses (connects) terminals T_1 and T_3; S_3 uses (connects) terminals T_2 and T_3; and finally, S_4 uses (connects) terminals T_3 and T_4. All terminals run at a fixed cost of, say, 5 for each terminal. All services generate a revenue, say, S_1 generates a revenue of 2, S_2 of 8, S_3 of 4, and finally, S_4 generates a revenue of 6. So the total revenue is 20, which breaks even with the total cost of running the terminals. The problem is how to allocate the total cost of the terminals on the services.

The way services use terminals can be modeled as bipartite graph. Try to argue why we can model the entire cost-sharing problem as the max-flow problem illustrated below. Find some max-flows, and argue for relevant solutions to the cost-sharing problem. Compare with standard cost-sharing methods based on equality and proportionality.

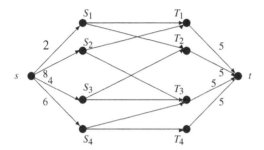

Try to formulate the problem as a linear programming problem, and comment on both its primary and dual formulations.

4.13. (Moulin and Sethuraman, 2013) A version of the max-flow problem relates to bipartite graphs with multiple source and sink nodes. Try to model a multiple-resource rationing problem, where agents have arbitrary claims over the resource types, as a max-flow bipartite graph. Suggest some ways to solve the problem. Comment on the connection with exercise 4.12.

References

Bergantiños, G., J. Gonzalez-Diaz, A. M. Gonzalez-Rueda, and M. P. Fernandez de Cordoba (2017), "Loss Allocation in Energy Transmission Networks," *Games and Economic Behavior* 102: 69–97.

Bogomolnaia, A., R. Holzman, and H. Moulin (2010), "Sharing the Cost of a Capacity Network," *Mathematics of Operations Research* 35: 173–192.

Ford, L. R., and D. R. Fulkerson (1962), *Flows in Networks.* Princeton, NJ: Princeton University Press.

Henriet, D., and H. Moulin (1996), "Traffic-Based Cost Allocation in a Network," *RAND Journal of Economics* 27: 332–345.

Hougaard, J. L., and H. Moulin (2014), "Sharing the Cost of Redundant Items," *Games and Economic Behavior* 87: 339–352.

Hougaard and Moulin (2018) Economic Theory, 65: 663–679.

Hougaard, J. L., J. D. Moreno-Ternero, M. Tvede, and L. P. Østerdal (2017), "Sharing the Proceeds from a Hierarchical Venture," *Games and Economic Behavior* 102: 98–110.

Hougaard, J. L., and M. Tvede (2015), "Minimum Cost Connection Networks: Truth-Telling and Implementation," *Journal of Economic Theory* 157: 76–99.

Jackson, M. O. (2005), "Allocation Rules for Network Games," *Games and Economic Behavior* 51: 128–154.

Jackson, M. O. (2008), *Social and Economic Network.* Newark, NJ: Princeton University Press.

Jackson, M. O., and A. Wolinsky (1996), "A Strategic Model of Social and Economic Networks," *Journal of Economic Theory* 71: 44–74.

Kalai, E., and E. Zemel (1982a), "Totally Balanced Games and Games of Flow," *Mathematics of Operations Research* 7: 476–478.

Kalai, E., and E. Zemel (1982b), "Generalized Network Problems Yielding Totally Balanced Games," *Operations Research* 30: 998–1008.

Kozen, D. C. (1992), *The Design and Analysis of Algorithms.* New York: Springer.

Meggido, N. (1974), "Optimal Flows in Networks with Multiple Sources and Sinks," *Mathematical Programming* 7: 97–107.

Moulin, H., and J. Sethuraman (2013), "The Bipartite Rationing Problem," *Operations Research* 61: 1087–1100.

Navarro, N. (2007), "Fair Allocation in Networks with Externalities," *Games and Economic Behavior* 58: 345–364.

Navarro, N. (2010), "Flexible Network Rules for Identified Externalities," *Games and Economic Behavior* 69: 401–410.

Reijnierse, H., M. Maschler, J. Potters, and S. Tijs (1996), "Simple Flow Games, *Games and Economic Behavior* 16: 238–260.

Rhys, J. M. W. (1970), "A Selection Problem of Shared Fixed Costs and Network Flows," *Management Science* 17: 200–207.

Sun X., and Q. Fang (2007), "Core Stability and Flow Games," in J. Akiyama et al. (eds.), *Discrete Geometry, Combinatorics and Graph Theory, vol. 4381, Lecture Notes in Computer Science.* New York: Springer.

5 Allocation in Decentralized Networks

We now turn to situations where a social planner (network manager) is unable to enforce a centralized solution, and agents may fail to coordinate when using the network or when interacting and forming connections.

For instance, in case of the Internet, no single authority has full control over network design and operations. Instead the Internet is operated by multiple competitive and autonomous entities, each seeking to optimize some individual performance objective, like profit, cost, throughput, flow, or delay. In such a context, inefficiencies are likely to occur. For instance, in the presence of congestion externalities, inefficiency can arise even from increased competition among service providers (Acemoglu and Ozdaglar, 2007). It is therefore important to study the extent to which a planner can influence the result of "anarchy" by intelligent use of potential instruments, such as deploying markets or protocols in the form of (nonmarket) allocation rules, distributing costs or benefits among network users.

In particular, we examine the influence of allocation rules on the efficiency of networks serving selfish (strategic) users, for instance, when designing large computer networks where autonomous agents aim at minimizing their own connectivity costs. We will consider the concept of agents rather broadly: for instance, as Internet service providers (ISPs) or computational agents optimizing some performance metric, given the actions taken by other agents (like nodes in a wireless ad hoc network programmed to act independently in response to actions taken by other nodes in the network).

In a broad class of problems, dubbed "network cost-sharing games," agents choose connection paths strategically in order to minimize their connectivity cost, knowing that the cost of every edge is shared among its users according to some preannounced allocation rule (protocol). The total cost of networks arising from pure-strategy Nash equilibria in the induced (noncooperative) game can be compared to a minimum-cost connection network using the metrics *price of anarchy* (PoA) and *price of stability* (PoS). For a given game, these measures are defined as the ratio between maximum social welfare, and the welfare in the worst and best Nash equilibria, respectively. Seminal results in Anshelevich et al. (2008) show that splitting edge costs equally among the edge users (dubbed the Shapley protocol) is a good allocation mechanism for inducing strategic behavior, often leading to near-optimal equilibria. Among the entire class of network cost-sharing games, worst-case PoA is a factor m (where m is the number of agents), and worst-case PoS is a factor $\mathcal{H}(m)$ (i.e., the mth harmonic number). For directed networks, this is actually the best we can hope to obtain with any allocation rule (ensuring existence of stable networks). For undirected networks, we can find allocation rules that improve efficiency considerably, but then we have to accept allocations that are far less compelling in terms of fairness.

However, networks need not be fixed physical distribution systems with connections in the form of wires, pipes, cables, or the like. Networks can also model agents' interactions of both social and economic nature: for instance, in communication and trade. These interactions determine the final network configuration, and when a societal value can be assigned to each such configuration, it is natural to examine how this value should be fairly allocated, as we did in section 4.5. In the present chapter, we will further study whether applied allocation rules are aligning private with social incentives. A seminal result in Jackson and Wolinsky (1996) shows that there is a tension between using a normatively compelling allocation rule and ensuring stability of a socially optimal network configuration. The severity of this tension is context dependent, and we will try to find some insights into how deep it goes. For instance, using the equal-split rule, we can always ensure that the efficient network configuration is stable, but then we have to accept that some components in the network may subsidize other components, even when no externalities are present between network components.

We will also explore noncooperative bidding mechanisms, which implement the efficient network configuration in every equilibrium outcome of the game and generate final payoffs corresponding to normatively desirable allocation rules, like the Shapley value. In particular, a certain bargaining game implements a MCST with final payoffs in the form of the Folk solution. None of these mechanisms require the existence of a social planner (or trusted third party) to whom the agents report cost or value information.

Finally, we consider a situation where the cost of servicing any coalition of users is concave, and every user has a private valuation of service. The planner (network administrator) wants to maximize system welfare but is unaware of users' valuations. To elicit these valuations, the planner can use a so-called Moulin mechanism that can be seen as an iterative ascending auction. The mechanism determines who gets service and at what price. If payment has to be budget balanced and the payment rule is population monotonic, the Moulin mechanism is group strategyproof (i.e., no group of users can game the system by misreporting their valuation). However, the mechanism may result in inefficiency. In particular, using the Shapley value as a payment rule minimizes the worst-case efficiency loss. Two applications (multicasting and bandwidth on demand) are briefly described.

It should be emphasized that the chapter is not intended to review the large literature concerning resource/quality-of-service allocation and pricing of services in decentralized networks. Not that this literature is irrelevant, but its primary focus concerns revenue maximization of service providers subject to users' strategic interaction. The problem is challenging, since prices, on top of being directly associated with marketing decisions of the service provider, also have implications for traffic management and network resource utilization, (see, e.g., the surveys in DaSilva, 2000; Altman et al., 2006; and Ozdaglar and Srikant, 2007).

5.1   Anarchy

The economics literature abounds with examples where private optimal behavior is in conflict with social interests. In situations where agents are autonomous and maximize their own individual objectives, a key question is therefore how to design networks, and allocation rules, such that the potential inefficiency of equilibrium behavior is minimized: the natural benchmark being social optimality that could be reached by a coordinated centralized solution. In the context of a routing game, the following iconic network example illustrates that there is a price to pay for "anarchy" in the sense of uncoordinated individual optimization.

Example 5.1 (Braess's paradox)   Consider 4,000 cars going from location s to t. The travel time is given in the graph below: two arcs (roads sB and At) have a fixed travel time of 45 minutes; two arcs (roads sA and Bt) have a travel time that depends on the number x of cars using this arc.

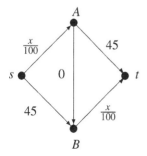

If a social planner is allowed to coordinate the cars in order to minimize the average travel time, we get the following optimal solution: let 2,000 cars go along the upper path, $s \to A \to t$, and 2,000 cars along the lower path, $s \to B \to t$. For each car, the total travel time will therefore be $2,000/100 + 45 = 65$ minutes. If the decision is left to the drivers themselves, this will actually also be the unique Nash equilibrium if there is no option of jumping from A to B.

Now, with the option of jumping from A to B, each driver will think as follows. Starting at s, the worst possible travel time at arc sA will be $4,000/100 = 40$ minutes, which is lower than the 45 minutes going along sB, so sA is chosen. Then, at A, the possibility of jumping to B will lead to the same line of reasoning: since Bt at worst takes 40 minutes, while At takes 45 minutes, each driver will choose to jump from A to B and go along Bt. Since every driver will think like this, each driver will end up getting a total travel time of $40 + 40 = 80$ minutes. Compared to the social optimum, with an average travel time of 65 minutes, this is clearly inefficient. So introducing the possibility of jumping from A to B, which at first glance seems to represent an improvement for the individual driver, actually leads to a bad equilibrium, where every driver is worse off than if jumping from A to B was impossible: hence the paradox.

A straightforward way to assess the price of selfish behavior is to compute the ratio between the equilibrium travel time and the socially optimal travel time. So in the case above, we see that the PoA is a factor $80/65 = 1.23$ compared to the socially optimal travel time. But the relative difference from the socially optimal travel time is even bigger when reducing the driving time on the roads sB and At to 40 minutes (the point of indifference between jumping from A to B or continuing): in this case the PoA becomes a factor $80/60 = 4/3 = 1.33$. In fact, this turns out to be the upper bound on the PoA in such routing games when arc costs grow linearly with traffic (see, e.g., Roughgarden, 2007).

The next section formalizes the notion of PoA and adds another useful inefficiency metric, the PoS.

5.1.1   PoA/PoS

The term "price of anarchy" (PoA) was originally coined in Papadimitriou (2001), and since then the concept has generated a sizable literature in computer science and game theory. As illustrated in example 5.1 above, the idea behind PoA is straightforward (Koutsoupias and Papadimitriou, 1999): for a given game, look at the set of equilibria, and compute the ratio between the worst objective function value of an equilibrium and that of a socially optimal outcome.

Formally, for a given game, consider the set of strategy profiles $\mathcal{S}$ with $\mathcal{NE} \subset \mathcal{S}$ being the (nonempty) set of Nash equilibria. Let $\mathcal{X}$ be a set of outcomes, and $x(s) \in \mathcal{X}$ the outcome induced by strategy profile $s \in \mathcal{S}$. Moreover, let $W : \mathcal{X} \to \mathbb{R}_+$ be a social welfare function. Then

$$\text{PoA} = \frac{\max_{s \in \mathcal{S}} W(x(s))}{\min_{s \in \mathcal{NE}} W(x(s))} \geq 1. \tag{5.1}$$

Similarly, considering a common cost function $C : \mathcal{X} \to \mathbb{R}_+$, we have

$$\text{PoA} = \frac{\max_{s \in \mathcal{NE}} C(x(s))}{\min_{s \in \mathcal{S}} C(x(s))} \geq 1. \tag{5.2}$$

We clearly want to minimize PoA, since this minimizes the relative difference between worst-case equilibrium performance and the result of a centrally coordinated outcome. For an entire class of games, we can speak of *worst-case* PoA, meaning the highest PoA among games in the class. Potentially, PoA can be unbounded, but surprisingly many classes of games have bounded PoA, for instance, the routing games with linear costs mentioned in example 5.1. So, for many problems, there is actually a limit to the efficiency loss resulting from decentralized decision making.

Instead of using the metric with respect to worst-case equilibrium behavior, we can focus on *best-case* equilibrium behavior, termed the PoS (see, e.g., Chen et al., 2010):

$$\text{PoS} = \frac{\max_{s \in \mathcal{S}} W(x(s))}{\max_{s \in \mathcal{NE}} W(x(s))} \geq 1. \tag{5.3}$$

Similarly, considering a common cost function $C : \mathcal{X} \to \mathbf{R}_+$, we have

$$\text{PoS} = \frac{\min_{s \in \mathcal{NE}} C(x(s))}{\min_{s \in \mathcal{S}} C(x(s))} \geq 1. \tag{5.4}$$

For every game, we have by definition PoA $\geq$ PoS ≥ 1. In some games, like Braess's paradox, PoA = PoS, but we will also encounter games where, for instance, PoS = 1, while PoA is unbounded.

5.2   Network Cost-Sharing Games

To focus on the role of allocation rules in inducing good equilibrium behavior, we consider a class of games known in the computer science literature as *network cost-sharing games* (see, e.g., Chen et al., 2010). These games resemble the situation in the connection networks model from section 4.4.

A network cost-sharing game consists of a set of agents $M = \{1, \ldots, m\}$ and a graph $G = (N, E)$, which can be directed or undirected. Every edge is associated with a constant cost $c_e \geq 0$. Every agent $i \in M$ wants to connect a pair of nodes (a_i, b_i). If G is a digraph, a_i is the source, and b_i the sink. Costs are allocated using a cost-additive rule. That is, for every edge, the cost c_e is shared using an edge-specific allocation rule (dubbed a *protocol*), that is, a mapping $y^e : 2^M \to \mathbb{R}_+^m$ assigning nonnegative cost shares to the set of agents using this edge. Agent i's total payment is then found by adding up i's cost share of all the edges used by i (see below).

The *strategy set* of agent i, $\mathcal{P}_i$, consists of all paths in G connecting (a_i, b_i). A strategy (path) profile $P = (P_1, \ldots, P_m)$, with $P_i \in \mathcal{P}_i$, for every $i \in M$, is an *outcome* of the game. Given outcome P, the resulting network becomes $\cup_i P_i$. The cost of P is simply the total cost of the resulting network, that is, $C(P) = \sum_{e \in \cup_i P_i} c_e$.

Given the cost-sharing protocol y^e, the cost of agent $i \in M$, induced by P, is given by

$$c_i(P) = \sum_{e \in P_i} y_i^e(S_e), \tag{5.5}$$

where $S_e = \{j \in M \mid e \in P_j\}$ is the set of agents using the edge e. The protocol y^e is budget balanced in the sense that $\sum_{i \in M} \sum_{e \in \cup_i P_i} y_i^e = C(P)$, where for every set S_e, $\sum_{i \in S} y_i^e(S_e) = c_e$: that is, agents do not pay for edges they do not use. Thus, $C(P) = \sum_{i \in M} c_i(P)$.

Agents choose strategies minimizing c_i; an outcome P is a pure strategy Nash equilibrium if no agent can reduce their cost c_i by deviating from P_i given $\{P_j\}_{j \neq i}$.

If a network cost-sharing game has a nonempty set of (pure strategy) Nash equilibria with resulting networks, we can compare these to the cost-minimizing network (satisfying all connection demands) using the indices PoA and PoS. In particular, it is natural to ask: What type of cost-sharing protocol ensures the existence of pure-strategy Nash equilibria for every game (i.e., induces stability) and minimizes worst-case PoA and/or PoS?

Suppose we use a protocol where the cost of each edge is shared equally among its users (dubbed the Shapley protocol):

$$y_i^e(S^e) = \frac{c_e}{|S_e|} \quad \text{for every} \quad i \in M. \tag{5.6}$$

With this protocol, network cost-sharing games have a finite set of strategies, and the cost of each agent depends only on the number of agents choosing each strategy. Consequently, network cost-sharing games based on the Shapley protocol belong to the class of *congestion games* analyzed in Rosenthal (1973): every such game possesses at least one pure-strategy Nash equilibrium.

Since the Shapley protocol seems compelling in terms of fairness, the interesting question is how well it performs in terms of efficiency. In other words, how close does the cost of resulting equilibria networks come to a minimum-cost connection network as measured by worst-case PoA and PoS? The following example illustrates.

Example 5.2 (Worst-case PoA)   Consider the (undirected) graph below, where two nodes s and t are connected via two parallel edges: the upper has cost 1; the lower has cost m. Here m agents have connection demand (s, t): their strategy consists of choosing the upper or lower edge to obtain connectivity.

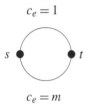

Clearly, the social planner will connect everybody via the upper edge with cost 1. However, using the Shapley protocol (5.6), selfish equilibrium behavior can be as bad as every agent choosing the lower edge. Indeed, if m agents have chosen the lower edge, the corresponding cost for every agent is $c_i = m/m = 1$, and consequently, every agent is indifferent between deviating by single-handedly choosing the upper edge with cost 1, and playing the equilibrium strategy with cost 1 as well. So the strategy profile where every agent chooses the lower edge is a pure-strategy Nash equilibrium in the game. Computing the PoA, we get $PoA = m/1 = m$. Notice that it does not matter whether the graph is directed or undirected in this case. As argued in Anshelevich et al. (2008), this is, in fact, the worst-case PoA of any network cost-sharing game using the Shapley protocol.

Since the PoA can be very high, as illustrated above, it is natural to examine whether the Shapley protocol can do better in terms of PoS. For instance, in example 5.2 above, everybody choosing the upper edge is also a pure-strategy Nash equilibrium, so $PoS = 1$ in this case. The following example, from Anshelevich et al., (2008), illustrates the bound on PoS when the graph G is directed.

Example 5.3 (Worst-case PoS)   Consider the digraph below, in which every agent $i \in M$ wants to connect a common source s to their respective sinks t_i. Arc costs are given as in the figure: every agent's direct connection st_i costs $1/i$; the cost of arc sz is $1 + \varepsilon$, and from z it is free to connect to individual sinks.

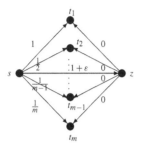

The socially optimal solution connects all agents via the common arc sz with total network cost $1 + \varepsilon$. However, using the Shapley protocol, this is not a Nash equilibrium, since agent m pays ε/m more than if she chose her direct connection $s \rightarrow t_m$ with cost $1/m$. If agent m deviates, so will agent $m - 1$, since now he pays $\varepsilon/(m - 1)$ more than if he chose his direct link, and so forth. The unique Nash equilibrium is therefore that every agent chooses their own direct link, st_i, with total cost being the mth harmonic number $\mathcal{H}(m) = \sum_{i=1}^{m} \frac{i}{m} \le 1 + \ln m$. As we will see below, this is precisely the worst-case PoS for any (directed) network cost-sharing game using the Shapley protocol.

Formalizing this example, Anshelevich et al. (2008) prove the following result.

Theorem 5.1   Using the Shapley protocol, the worst-case PoS of any (directed) network cost-sharing game is $\mathcal{H}(m) = 1 + \frac{1}{2} + \cdots + \frac{1}{m}$.

Proof [Sketch of argument]   By Rosenthal's existence result, a network cost-sharing game with Shapley protocol is an *exact potential game* (Monderer and Shapley, 1996)[1] with potential

$$B(P) = \sum_{e \in \cup_i P_i} \sum_{x=1}^{|S_e|} \frac{c_e}{x},$$

for each strategy vector P.

---

1. A game is called an exact potential game if it admits a potential (i.e., a function) $B : \mathcal{P}_1 \times \cdots \times \mathcal{P}_m \rightarrow \mathbb{R}$ such that, for every $i \in M$, and pair of outcomes, P and $P' = (P_i', P_{-i})$, we have $B(P') - B(P) = c_i(P') - c_i(P)$.

Let P^* be the strategy vector defining the socially optimal solution. Then

$$B(P^*) = \sum_{e \in \cup_i P_i^*} \sum_{x=1}^{|S_e|} \frac{c_e}{x} = \sum_{e \in \cup_i P_i^*} c_e \mathcal{H}(|S_e|)$$

$$= C(P^*)\mathcal{H}(|S^e|) \le C(P^*)\mathcal{H}(m).$$

Now, pure-strategy Nash equilibria, $\overline{P}$, are minimizers of the potential, so $B(\overline{P}) \le B(P^*)$. Moreover, for any strategy vector P, we have $B(P) \ge C(P)$. Thus, there exists a Nash equilibrium with cost at most $C(P^*)\mathcal{H}(m)$. ∎

While worst-case PoS is precisely $\mathcal{H}(m)$ when G is a digraph, it is *at most* $\mathcal{H}(m)$ when G is undirected.[2]

Chen et al. (2010) ask whether there exist cost-sharing protocols,[3] guaranteeing the existence of pure-strategy Nash equilibria,[4] that do better in terms of efficiency than the Shapley protocol. For *directed* networks, the answer is negative. Indeed, consider PoA.

Let $(P_1^*, \ldots, P_m^*)$ be an optimal outcome with total cost C^*. By choosing P_i^*, agent i can ensure a payment no greater than C^* (independent of the choice of protocol and the other agents' strategies). Thus, by budget balance, the cost of any pure-strategy Nash equilibrium must be at most mC^*. To see that this can indeed be the case, consider the graph below (which is a variation of example 5.3 above).

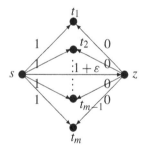

Clearly, the optimal cost is $C^* = 1 + \varepsilon$.

---

2. No tight lower bound seems to be known when G is undirected. However, for games where every agent is a node wanting to connect to a single common source, Bilò et al. (2014) show that worst-case PoS is constant.

3. That is, methods assigning a budget-balanced cost-sharing rule y^e to each edge, where y^e is function only of the edge cost c_e, and the set of agents M.

4. In addition to Chen et al. (2010), see Gopalakrishnan et al. (2014) for general conditions on protocols guaranteeing the existence of pure-strategy Nash equilibria. Basically, the protocols should be equivalent to generalized weighted Shapley value rules.

The outcome $P_i = st_i$, for every agent $i \in M$, is a Nash equilibrium independent of the choice of protocol, and it has cost mC^* for $\varepsilon \to 0$. Thus, the worst-case PoA of any protocol in m-agent directed networks is m. So in this sense the Shapley protocol is the best we can do in terms of worst-case PoA.

Concerning worst-case PoS, we need to elaborate on the case from example 5.3 above. I claim that, in this example, $P_i = st_i$, for every $i \in M$ and with cost $\mathcal{H}(m)$, is the unique Nash equilibrium for any choice of protocol y^e. Indeed, it can be shown that protocols are monotone in the sense that $y_i^e(S_e) \geq y_i^e(T_s)$ when $i \in S_e \subseteq T_e \subseteq M$. Thus, by budget balance, $y_j^e(\{1, \ldots, j\}) \geq 1/j$ for each $j = m, \ldots, 1$. Now consider the graph in example 5.3. Let S_{sz} be the set of agents choosing the arc sz. Suppose that there is an equilibrium for which $S_{sz} \neq \emptyset$. Let j be the agent with highest index in S_{sz}. Since any protocol is monotone, $y_j^{sz}(S_{sz}) \geq y_j^{sz}(\{1, \ldots, j\}) \geq (1+\varepsilon)/j$, but this contradicts the assumption that $S_{sz} \neq \emptyset$ in equilibrium, since j can reduce its cost by choosing the direct link st_j with cost $1/j$.

Thus, the worst-case PoS of every protocol in m-agent directed networks is at least $\mathcal{H}(m)$. So in this sense, the Shapley protocol is the best we can do in terms of worst-case PoS.

Now, consider undirected networks. Here, Chen et al. (2010) show that so-called *ordered protocols* perform much better than the Shapley protocol in terms of worst-case PoA. Based on an ordering π of the agents, ordered protocols make the first agent pay the full cost of all edges in his path; the second agent then pays the full cost of every edge in her path not already paid for, and so forth. That is, $y_i^e(S_e) = c_e$ if $\pi(i) \leq \pi(j)$ for all $j \in S_e$, and is 0 otherwise.

Consider, for instance, an ordering of the agents based on the following Prim-like procedure in a situation where the agents want to connect individual sources to a common sink. Let the first agent, say, i, be the one with the smallest path cost to the common sink; let the second agent be the one with smallest path cost to either the common sink or the source of the first agent s_i; and so forth. An outcome based on this Prim protocol is a pure-strategy Nash equilibrium, because given the ordering and cost shared using an ordered protocol, no agent can reduce their cost by deviating. Indeed, the first agent pays his minimum cost, the second agent pays her minimum cost given the choice of the first agent, and so on.

In Chen et al. (2010) it is shown that such a Prim protocol has a PoA of at most 2, which can be compared to worst-case PoA of the Shapley protocol being m, even in networks with parallel edges, as in this case.

However, from the viewpoint of fairness, ordered protocols are obviously less compelling than the Shapley protocol, so improved efficiency comes at the price of unfair division of costs.

5.2.1 Strong Nash Equilibrium

To address the fact that agents may be able to coordinate their actions, we can strengthen the equilibrium concept from pure Nash to *strong Nash equilibrium* (i.e., outcomes for which no coalition of agents can deviate and reduce the cost of all its members). While the concept of strong Nash equilibrium is more compelling in terms of stability, many games do not admit such equilibria. For instance, consider the following example from Epstein et al. (2009).

Example 5.4 (No strong Nash equilibrium) Consider the following digraph, where Ann wants to connect (s_A, t_A), and Bob wants to connect (s_B, t_B).

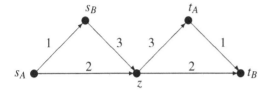

Using the Shapley protocol, there is a unique Nash equilibrium given by $P_A = \{s_A z, z t_A\}$ and $P_B = \{s_B z, z t_B\}$ with costs $c_A(P) = c_B(P) = 5$. However, through a coordinated deviation P' with $P'_A = \{s_A s_B, s_B z, z t_A\}$ and $P'_B = \{s_B z, z t_A, t_A t_B\}$, they both get a reduced cost $c_A(P') = c_B(P') = 4$. Thus, the unique Nash equilibrium is not a strong Nash equilibrium.

Epstein et al. (2009) provide sufficient conditions on the network structure for existence of strong Nash equilibria in network cost-sharing games with directed graphs and Shapley protocol (see their theorem 3.5). In terms of network design, it is interesting to note that what determines the existence of stable outcomes is network topology, and not the specific cost of edges.[5]

Moreover, Epstein et al. (2009) prove the following result.

Theorem 5.2 Consider network cost-sharing games with directed graphs and Shapley protocol possessing strong Nash equilibria. The (strong) PoA of such games is at most $\mathcal{H}(m)$.

Proof Following Epstein et al. (2009), let $\widehat{P}$ be a strong Nash equilibrium and P^* be the optimal outcome. Consider an ordering of the agents based on a sequence of coalitions:

$$T_1 = \{1\}, \ldots, T_k = \{1, \ldots, k\}, \ldots, T_m = M.$$

---

5. See, e.g., Holzman and Law-yone (1997, 2003) for related results.

The order is created based on the fact that since $\widehat{P}$ is a strong Nash equilibrium, there exists an agent $j \in T_j$ such that

$$c_j(\widehat{P}) \leq c_j(P_{T_j}^*, \widehat{P}_{M \setminus T_j}).$$

Let $c_j(P_{T_j})$ be the cost of agent j for outcome P in the projected game with agent set T_j. Since we are applying the Shapley protocol and all the games are potential games, we get

$$c_j(P_{T_j}^*, \widehat{P}_{M \setminus T_j}) \leq c_j(P_{T_j}^*) = B(P_{T_j}^*) - B(P_{T_{j-1}}^*).$$

Thus, summing over all agents,

$$\sum_{i \in M} c_i \leq B(P_{T_m}^*) = B(P^*) = \sum_{e \in \cup_i P_i^*} c_e \mathcal{H}(S_e) \leq C(P^*)\mathcal{H}(m). \qquad \blacksquare$$

Recall the digraph in example 5.3: in that case the unique pure-strategy Nash equilibrium, $\overline{P}_i = st_i$, for every i, is a strong Nash equilibrium. This example therefore illustrates the worst-case PoA in strong Nash equilibrium.

Since the set of strong Nash equilibria is contained by the set of pure-strategy Nash equilibria, it follows that (strong) PoA $\leq$ PoA, but recalling that the worst-case PoA for the Shapley protocol is m, theorem 5.2 shows that we can really do much better with strong Nash.

Moreover, by the same argument, (strong) PoS $\geq$ PoS, and since (strong) PoA $\geq$ (strong) PoS, it follows directly from theorem 5.2 that (strong) PoS is at most $\mathcal{H}(m)$.

5.3   Network Formation Games

Strategic decisions not only influence the use of existing networks but may also shape the network itself. We now turn to networks formed by agents with full discretion to create, maintain, or sever connections, as they prefer. Following the seminal model of Jackson and Wolinsky (1996), we consider agents that interact, and these interactions are modeled by specific network configurations with edges representing pairwise relationships. Each network configuration has a societal value, which is distributed to individual agents by some allocation rule. Individual agents decide whether to form pairwise connections or to sever connections, based on the net benefit of such (inter)actions.

Recall the network games model from section 4.5. In that model, agents $i \in N$ are identified by nodes, and every (undirected) network $G \in \mathcal{G}$ connecting agents has a societal value given by a value function $v : \mathcal{G} \to \mathbb{R}$. A network is efficient if it has maximum value. The value of a network is allocated among individual agents using a budget-balanced allocation rule $\phi : \mathcal{G} \times V \to \mathbb{R}$. In this sense, $\phi_i(G, v)$ represents the net benefit (i.e., benefit minus cost) that agent i obtains if the network G is realized, under value function v.

In particular, let $Z(G)$ be the set of components of G. Then v is component additive if $v(G) = \sum_{G' \in Z(G)} v(G')$. Moreover, letting $H(G) = \{i \in N \mid \exists j : ij \in G\}$ be the set of active agents in G, the allocation rule ϕ is said to be component balanced if $\sum_{i \in H(G')} \phi_i(G, v) = v(G')$, for every G, and $G' \in Z(G)$, and component-additive v.

Since agents have full discretion to form connections it seems natural that ϕ induces a noncooperative game, where every agent wants to maximize her net benefit (payoff) by choosing a set of other agents with whom to connect, and connections are realized between agents naming each other. Formally, the strategy set of every agent $i \in N$ is given by $S_i = 2^{N \setminus \{i\}}$. For a given strategy profile $s \in \times_{i \in N} S_i$, the network $G(s) = \{ij \mid j \in s_i, i \in s_j\}$ is realized, resulting in the payoff $\phi_i(G(s), v)$ to every agent $i \in N$. Denote the game by

$$(N, \{S_i\}_{i \in N}, \phi). \tag{5.7}$$

A standard equilibrium concept like pure-strategy Nash equilibrium can be applied and exists for all games. Indeed, $s_i = \{\emptyset\}$ for all i, leading to the empty network $G = \{\emptyset\}$, is a pure-strategy Nash equilibrium for every game (5.7) and can be so even in undominated strategies. Thus, PoA is unbounded for every such game.

However, the unilateral deviations from the Nash equilibrium notion do not take into consideration that it takes two agents to agree on forming a connection: the notion only guarantees that a given agent can never gain by severing a connection. Therefore, Jackson and Wolinsky (1996) search for alternative notions of stability and suggest focusing on pairwise stability.

5.3.1 Pairwise Stability

A network $G \in \mathcal{G}$ is *pairwise stable* if

1. for all $ij \in G$, we have $\phi_i(G, v) \geq \phi_i(G - ij, v)$, and $\phi_j(G, v) \geq \phi_j(G - ij, v)$, and
2. for all $ij \notin G$, if $\phi_i(G + ij, v) > \phi_i(G, v)$, then $\phi_j(G + ij, v) < \phi_j(G, v)$.

In words, a network is pairwise stable if no agent wants to sever a connection, and no pair of agents can agree on adding a connection.

Pairwise-stable networks may not exist, as demonstrated by the following example from Jackson (2008).[6]

Example 5.5 (No pairwise-stable network)   Consider a situation with four nodes (agents). Establishing a connection between any pair of nodes *ij* costs 5 for each of the two agents involved. For each agent, the benefit is 0 being alone, 12 being connected to one agent, 16 being connected to two agents, and 18 being connected to three agents. Assume that the allocation rule ϕ assigns each agent her net benefit (benefit minus cost) with the value of a network being the total sum of individual net benefits.

Notice that an efficient network connects all four agents via three links, as shown below (where the numbers indicate the net benefit of the agent) with a total net benefit (welfare) of $2(18 - 5) + 2(18 - 10) = 42$.

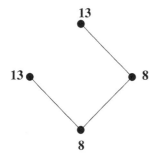

The problem is that the agents are symmetric in terms of costs and benefits but are treated differently by the efficient network configuration: Who should get 13, and who should get 8? Each of the low-payoff agents has incentive to sever the connection to a high-payoff agent, because this generates a net benefit of $16 - 5 = 11$ (since she is now connected to two agents, but pays only for one connection), leading to the network below.

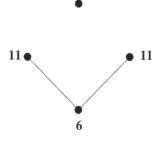

6. See also Jackson and Watts (2002).

But here the problem remains: the low-payoff agent has incentive to sever a link to a high-payoff agent: getting net benefit $12 - 5 = 7$.

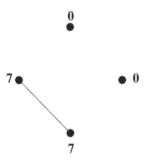

Clearly, the two singleton agents, with zero payoff, will now have an incentive to create a connection, so we get a network consisting of two components.

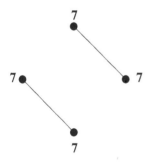

But in this case any pair of disconnected agents has an incentive to create a connection, gaining a net benefit $18 - 10 = 8$. Thus, we have created a cycle, since this results in the network we started out with, so no pairwise-stable network exists.

However, the graphs in both the third and fourth steps above represent a Nash equilibrium (different from the empty graph) in the noncooperative connection game (5.7). In fact, this illustrates the difference from the Nash equilibrium notion: the first and second graphs cannot be the result of a Nash equilibrium, since there exists an agent with a profitable deviation based on a unilateral move to sever a connection; in the third and fourth graphs, it takes two agents to make the coordinated bilateral action of naming each other.

For many allocation rules, pairwise-stable networks are guaranteed to exist. For instance, it is intuitively clear that if the societal value of a network is split equally among the agents, then nobody can improve on an efficient graph, and consequently, efficient networks will be pairwise stable. But more sophisticated allocation rules also can guarantee the existence of pairwise-stable networks. Recall the definition of the extended Myerson value (4.29) from section 4.5.2. There exists a pairwise-stable network for every value function v when using the extended Myerson value ϕ^{EMV} (Jackson, 2003). In particular, this means that improving cycles, as in example 5.5 above, cannot occur (indeed, using ϕ^{EMV} as the allocation rule in example 5.5, the efficient graph is now pairwise stable, since a low-payoff agent will reduce her net benefit by 7 if she severs her connection to a high-payoff agent; see exercise 5.5).

Pairwise-stable networks may be inefficient if there are externalities; that is, if the net benefit of a given agent is influenced by connections formed by other agents. Consider the following example.

Example 5.6 (All pairwise-stable networks are inefficient)   Consider a situation with three nodes (agents). Establishing a connection between any pair of nodes ij costs 5 for each of the two agents involved. The benefit of each agent in a network of the shape G' below is 16, while it is 20 in the complete graph G''. The allocation rule ϕ assigns each agent her net benefit, and the value of the network is the total sum of net benefits if the network is of the form G' or G'', and $v(G) = 0$ otherwise. Thus, the situation is captured by the two graphs below, where numbers indicate the net benefits of the respective nodes.

The complete graph G'' is efficient with maximum value 30. However, G'' is not pairwise stable, since all agents have incentive to sever one of their two connections (thereby obtaining net benefit 11 versus 10 in the complete graph). The three possible graphs with configuration G' are inefficient, since the value is only 28, but unlike the efficient graph, they are pairwise stable.

Looking at the PoA with respect to pairwise stability we thus get PoA = $30/28 = 1.07$ in the present case (which also equals PoS).

It turns out that for a large class of well-behaved allocation rules, stability and efficiency do not necessarily go hand in hand. The following result from Jackson and Wolinsky (1996) highlights the tension between pairwise stability and efficiency.

Theorem 5.3 For $n \geq 3$, there does not exist an anonymous and component-balanced allocation rule ϕ, such that for each $v \in V$, at least one efficient network is pairwise stable.

Proof Following the argument in Jackson and Wolinsky (1996), we consider first the three-agent case $N = \{i, j, k\}$, where the component additive value function v is such that, for all i, j, k, $v(\{ij\}) = v(\{ij, jk, ik\}) = 1$, and $v(\{ij, jk\}) = 1 + \varepsilon$. Consequently, networks of the form $G = \{ij, jk\}$ are efficient. By anonymity and component balance, we have that $\phi_i(\{ij\}, v) = 0.5$ and $\phi_i(\{ij, jk, ik\}, v) = \phi_k(\{ij, jk, ik\}, v) = 0.33$. Now, pairwise stability of the efficient graph requires that $\phi_j(\{ij, jk\}, v) \geq \phi_j(\{ij\}, v) = 0.5$, implying that

$$\phi_i(\{ij, jk\}, v) = \phi_k(\{ij, jk\}, v) \leq 0.25 + 0.5\varepsilon.$$

But this contradicts pairwise stability, since (for sufficiently small ε) i and k can both gain by forming a connection (obtaining 0.33). For $n > 3$, assign $v(G) = 0$ to any graph G involving agents other than i, j, and k. ∎

Recall that the equal-split rule guarantees that the efficient network is pairwise stable, but the equal-split rule is not component balanced, so clearly component balance plays a key role in the tension between stability and efficiency highlighted by theorem 5.3. Now, if the value function v is component additive, then the component-wise equal-split rule (i.e., the rule that splits the value of every component equally amongs the members) is clearly component balanced and can still guarantee the existence of a pairwise-stable network (indeed, the following graph will be pairwise stable: take the biggest component that maximizes the component-wise equal-split payoff to its members, then do the same among the remaining nodes, and so forth. By construction, no member in any component wishes to sever a link, and since v is component additive, bridging two components will reduce the payoff of the members in the higher-payoff component, so no pair of agents, one in each of the two different components, will agree on forming a bridge). But by theorem 5.3 above, such a network need not be efficient.

5.3.2 Alternative Stability Notions

The notion of pairwise stability itself also plays a crucial role in the tension displayed in theorem 5.3. Jackson and Wolinsky (1996) suggest that it would be natural to strengthen pairwise stability by allowing for side payments, such that any pair of agents can agree on establishing a connection if it is *jointly* beneficial.

When allowing for side payments, we say that the network G' *defeats* G, under allocation rule ϕ and value function v, if either

$$G' = G - ij \quad \text{and} \quad \phi_i(G, v) < \phi_i(G', v) \text{ or } \phi_j(G, v) < \phi_j(G', v),$$

or

$$G' = G + ij \quad \text{and} \quad \phi_i(G', v) + \phi_j(G', v) > \phi_i(G, v) + \phi_j(G, v).$$

In other words, a network is defeated if any agent wants to sever a link, or if any pair of agents can increase their total payoff by forming a link (where one agent may need to compensate the other agent).

A network G is *pairwise stable allowing for side payments* if it is undefeated. Jackson and Wolinsky (1996) show that for $n \geq 3$, there is no component-balanced allocation rule ϕ *such that for each value function v, no efficient network is defeated by an inefficient one*. Since allowing for side payments strengthens of pairwise stability, it is not surprising that the tension is maintained with weaker conditions on the allocation rule: notice that compared to theorem 5.3, we no longer need anonymity of ϕ for the result to hold.

Following Dutta and Mutuswami (1997), we can also consider stability notions related to equilibria of the noncooperative game defined in (5.7). Clearly, these notions have to be stronger than the standard Nash equilibrium, as discussed above. A network is called *strongly stable* if it results from a strong Nash equilibrium of the game (5.7). A strongly stable network is not necessarily pairwise stable, because there may be two agents who want to form a connection between them with one being strictly better off and the other being indifferent (this coalition does not have an improvement in the sense of a strong Nash equilibrium, since both agents must be strictly better off by deviating). With this strong (Nash) stability notion, Dutta and Mutuswami (1997) are able to prove the following result.

Theorem 5.4 Let $n \geq 3$. There exists a component-balanced allocation rule ϕ, which is anonymous on the set of strongly stable networks, such that for every value function v (assigning strictly positive values to all networks except for the totally disconnected graph), the set of strongly stable networks is non-empty and contains only efficient networks.

From theorem 5.4, it appears that the tension between stability and efficiency is not as deep as expected, but it is difficult to compare theorems 5.3 and 5.4 directly, since strongly efficient networks are not necessarily pairwise stable. Moreover, other important differences exist: there is now a restriction on the class of value functions, and the allocation rule is only anonymous on the set of strongly stable networks and not in general. Both features are central for the result to hold. While the former restriction on the set of value functions seems fairly harmless, it is perhaps more difficult to accept that anonymity may be violated by the allocation rule outside the set of strongly stable networks. In any case, the theorem indicates that to overcome the tension between stability and efficiency, we have to either accept that agents can be treated differently or abolish the idea of component balance when designing our allocation rule. Since component balance must hold only when v is component additive (i.e., when there is no externality between the components), neither of the two issues at stake seems easy to give up.[7]

5.3.3 Directed Networks

Directed networks no longer require a bilateral agreement to form a connection. Indeed, in computer networks, information is often sent from one agent to another without the consent of the receiver (e.g., spam mail). Therefore, the naturally associated noncooperative game is now one where the agents simultaneously name other agents to whom they want a connection, and the resulting network is the union of these desired connections. Formally, agents' strategy sets are given by $S_i = 2^{N \setminus \{i\}}$. For a given strategy profile $s \in \times_{i \in N} S_i$, the network $G'(s) = \{ij \mid j \in s_i\}$ is realized, resulting in the payoff $\phi_i(G'(s), v)$ to every agent $i \in N$. Denote this game by

$$(N, \{S_i\}_{i \in N}, \phi). \tag{5.8}$$

Stability in directed networks is naturally represented by pure-strategy Nash equilibria of game (5.8). Contrary to game (5.7), the strategy $s_i = \{\emptyset\}$, for all i, is not an equilibrium in all games (5.8), but since creation of connections no longer requires mutual consent from both agents, it is easy to imagine inefficient equilibria, where some agents have a positive payoff from "spamming" other agents.[8]

Let $N(G) = \{i \mid \exists j : ij \in G \text{ or } ji \in G\}$ denote the set of active agents in G. Agent $i \in N$ is an *outsider* of G if $ij \in G$ for some $j \in N(G)$; $ki \notin G$ for all $k \in N(G)$; for every $j \neq i$, $j \in N(G)$, there exists $k \neq i$, $k \in N(G)$, such that

---

7. Further discussion and variations of theorems 5.3, and 5.4 can be found in Jackson (2008). See also exercise 5.7.

8. Bala and Goyal (2000) study the structure of Nash-equilibrium networks in the special case where every agent's payoff is a strictly increasing function of the number of agents to whom there is a connection either directly or indirectly (e.g., if any form of connection to another agent yields a constant benefit), and a strictly decreasing function of the number of direct connections (e.g., if a constant cost is incured for every direct connection to another agent). They show that equilibrium networks are either empty or minimally connected.

$kj \in G$. In words, an outsider i connects to some active agent, but no active agent connects to i, and every active agent, except for i, has some other active agent connecting to him. Thus, a network contains at most one outsider.

Let $G - i$ be the result of deleting all i's connections in G. Now, an allocation rule ϕ is *directed component-balanced* if it is component balanced, and for any component-additive v, digraph G, and outsider i, we have that $v(G) = v(G - i)$ implies $\phi(G, v) = \phi(G - i, v)$. Thus, if a component contains an outsider, i, then i does not contribute to the value of the component and consequently should not be allocated any value, either.

With these new notions, Dutta and Jackson (2003) show that there is a parallel result to theorem 5.3 in the context of directed networks.

Theorem 5.5   For $n \geq 3$, there is no anonymous and directed component-balanced allocation rule ϕ such that, for each v, at least one efficient network is the result of a pure-strategy Nash equilibrium in game (5.8).

Proof [Sketch of argument]   Consider $n = 3$ and digraphs $G = \{12, 23, 31\}$ and $G' = \{13, 32, 21\}$ with value $v(G) = v(G') = 1 + \varepsilon$. Let $v(G'') = 1$ for every $G'' \neq G, G'$. So G and G' are the only efficient graphs. By anonymity and budget balance of ϕ, we have $\phi_1(G, v) = \phi_2(G, v) = \phi_3(G, v) = \frac{1+\varepsilon}{3}$. Now, consider the digraph $G'' = \{12, 21\}$. By anonymity and component balance of ϕ, we have $\phi_1(G'', v) = \phi_2(G'', v) = 0.5$. Moreover, $\phi_1(G'' + 31, v) = \phi_2(G'' + 31, v) = 0.5$ by anonymity and directed component balance, since agent 3 is an outsider in $G'' + 31$. Thus, the strategy profile $s^* = \{2, 3, 1\}$ leading to G is not a Nash equilibrium for $\varepsilon < 1/6$, since agent 2 will sever the arc 23 and add the arc 21 instead, resulting in the digraph $G'' + 31$ with payoff 0.5. A similar argument applies for strategy profile $s' = \{3, 1, 2\}$ leading to G'.   ∎

5.4   Bidding Mechanisms

An efficient network maximizes the potential aggregate surplus that agents can obtain by forming connections with each other. In section 4.5, we considered how this surplus can be allocated among agents by extending solutions from cooperative game theory to so-called network (value) games. For example, assuming that any coalition of agents connect in configurations that maximize their aggregate value, the Shapley value of the derived game (N, w), as defined in (4.31), represents a compelling way to share the surplus (Jackson, 2005).

In a situation without a social planner to enforce a particular allocation rule, it is natural to ask whether a reasonable allocation of network surplus can be obtained via some kind of (noncooperative) bargaining process among the agents themselves.[9] In particular, we aim at mechanisms for which (all) equilibrium strategies result in formation of efficient networks.

Assuming that the value function is common knowledge, we follow Pérez-Castrillo and Wettstein (2001, 2005), and define a network-formation process as a bargaining game in which connected network components are formed sequentially: first, there is a bidding stage, where agents bid over who becomes a proposer; then, in a second stage, the selected proposer chooses a network and makes an offer to every other agent. If all agents accept the proposal, the chosen network is implemented and the proposer is allowed to keep its value, while having to pay what was offered to every other agent (if the network only involves a subset of the agents, those left out subsequently run their own game). If just one agent rejects, the proposal is rejected, and the game starts all over, but without the proposer.

Formally, the mechanism is defined recursively on the number n of agents. If $n = 1$, the agent is "born" as the proposer and can only form the empty graph, therefore obtaining value $v(\emptyset) = 0$. Given that the rules of the game are known for at most $n - 1$ agents, we can define the game for n agents, consisting of three stages:

1. Each agent $i \in N$ makes a bid $b_j^i \in \mathbb{R}$ for every agent $j \in N$, with $\sum_{j \in N} b_j^i = 0$. Agents bid simultaneously. For each $i \in N$, let $B_i = \sum_{j \in N} b_i^j$ be the aggregate bid to agent i, and choose, according to an arbitrary tie-breaking rule, a proposer $i^* \in \{\arg\max_i B_i\}$ with maximum aggregate bid. Before moving on to stage 2, every agent $i \in N$, pays $b_{i^*}^i$ and receives B_{i^*}/n.

2. The proposer i^* chooses a subset of agents $S_{i^*} \ni i^*$ and a connected graph $G^* \in G^{S_{i^*}}$, and offers payments $x_j^{i^*} \in \mathbb{R}$ to every other agent $j \in N \setminus \{i^*\}$.

3. Sequentially, the agents in $N \setminus \{i^*\}$ either accept or reject the proposal of i^*. If an agent rejects the proposal, it is off the table. Otherwise the proposal is accepted. If the proposal is accepted, the final payoff to agents $i \in S_{i^*} \setminus \{i^*\}$ is $x_i^{i^*} - b_{i^*}^i + B_{i^*}/n$. The proposer i^* gets $v(G^*) - \sum_{i \neq i^*} x_i^{i^*} - b_{i^*}^i + B_{i^*}/n$; agents $i \in N \setminus S_{i^*}$ gets $x_i^{i^*} - b_{i^*}^i + B_{i^*}/n$ plus their payoff in the reduced game played by $N \setminus S_{i^*}$. If the proposal is rejected, the payoff to i^* is $-b_{i^*}^{i^*} + B_{i^*}/n$; everybody else gets $-b_{i^*}^i + B_{i^*}/n$ plus their payoff in the reduced game played by $N \setminus \{i^*\}$.

Notice that running this game does not require a planner to whom the agents have to report (compare to the mechanisms in chapter 6).

---

9. Finding noncooperative foundations for cooperative solution concepts is known in the literature as the *Nash program*, see, e.g., Serrano (2008) for a survey.

Pérez-Castrillo and Wettstein (2005)[10] are now able to show that not only is an efficient network formed in every equilibrium outcome of the game, but the payoffs are also uniquely determined by the Shapley value of Jackson's[11] monotonic cover (N, w) of the network game (N, v), defined in (4.30), section 4.5.

Theorem 5.6 At any subgame perfect Nash equilibrium (SPNE) of the bidding mechanism, an efficient network is formed, and payoffs to the agents are uniquely determined by the Shapley value ϕ^{SW} (4.31) of the cooperative game (N, w) given by (4.30).

Proof [Sketch of argument] The theorem trivially holds for the case $n = 1$; assume that it holds for all $m \leq n - 1$. By induction, we need to show it holds for $m = n$.

Consider a network game (N, v) with n agents. The claim is that the following strategies are the unique SPNE of the mechanism:

1. Each agent $i \in N$ bids $b^i_j = \phi^{SW}_i(N \setminus \{j\}, w) - \phi^{SW}_i(N, w)$ for every $j \neq i$ and $b^i_i = w(N) - w(N \setminus \{i\}) - \phi^{SW}_i(N, w)$.

2. The proposer i^* chooses $S_{i*} \subseteq N$, and a graph G^* such that $v(G^*) + w(N \setminus S_{i*}) = w(N)$. Moreover, i^* offers $x^{i^*}_j = \phi^{SW}_j(N \setminus \{i^*\}, w)$ to every $j \in S_{i*} \setminus \{i^*\}$, and $x^{i^*}_j = \phi^{SW}_j(N \setminus \{i^*\}, w) - \phi^{SW}_j(N \setminus S_{i*}, w)$ to every $j \notin S_{i*}$.

3. If i^* is the proposer, agents $i \in S_{i*}$ accept any offer greater than or equal to $\phi^{SW}_i(N \setminus \{i^*\}), w)$ and reject otherwise. Agents $i \notin S_{i*}$ accept any offer greater than or equal to $\phi^{SW}_i(N \setminus \{i^*\}, w) - \phi^{SW}_i(N \setminus S_{i*}, w)$ and reject otherwise.

The above strategy is a SPNE: indeed, suppose that $S_{i*} = N$; by rejecting the proposal, agents $i \neq i^*$ obtain $\phi^{SW}_i(N \setminus \{i^*\}, w)$ (implied by the induction argument). Thus, i^* has to offer $x^{i^*}_i \geq \phi^{SW}_i(N \setminus \{i^*\}, w)$ in order to make i accept. Now, notice that $b^i_j = b^j_i$ for all i, j (implied by the fairness property of Myerson (1.23), see section 1.3.1). Thus, $B_i = 0$ for all i, so the final payoff to all agents $i \in N$ is equal to $\phi^{SW}_i(N, w)$. Consequently, i^* will not gain by *not* being the proposer (which will also result in the Shapley value payoff). An agent $i \neq i^*$ cannot gain by raising the bid to become the proposer either: i has to pay the full cost $b^{i^*}_i$ but only receives B_{i*}/n in compensation.

By the induction argument, all agents $i \neq i^*$ must behave as in stage 3 in any SPNE. Moreover, in any SPNE $B_i = 0$, for all i, and therefore final payoffs do

---

10. See also Mutuswami et al. (2004).

11. Jackson (2005).

not depend on who is chosen as proposer. The bids given in stage 1, are the only bids satisfying these requirements.

Finally, it is clear that the proposer maximizes her payoff by forming an efficient component, and thereby the efficient network will be formed as a final result of running the mechanism. ∎

In equilibrium, we will select an arbitrary proposer (since $B_i = 0$ for all $i \in N$), but it does not matter for the final payoff who is selected. Thus, we could alternatively consider a random selection of the proposer with equal probability, as in Hart and Mas-Colell (1996), but in their theorem 2 (with $\rho = 0$), equilibrium payoffs are given by the Shapley value in expectation.

A nice feature of the mechanism is that if an agent bids according to the unique equilibrium bidding strategy in stage 1 (i.e., as $b_j^i = \phi_i^{SW}(N \setminus \{j\}, w) - \phi_i^{SW}(N, w)$ for every $j \neq i$ and $b_i^i = w(N) - w(N \setminus \{i\}) - \phi_i^{SW}(N, w))$, her final payoff is at least equal to her Shapley value payoff, *independent* of the other agents' bids. Thus, agents can not manipulate the bidding stage, either individually or in groups.[12]

Slikker (2007) presents similar types of bidding mechanisms for which equilibrium payoffs coincide with the (extended) Myerson value, and the componentwise equal-split rule. Van den Brink et al. (2013) suggest a variation of the mechanism for which equilibrium payoffs coincide with the average tree (AT) solution.

5.4.1 Bargaining in Connection Networks

In the MCST model, the value of a spanning tree is given by its negative total cost, and the value of any other graph can be set to $-\infty$. Thus, a modification of the bidding mechanism in Pérez-Castrillo and Wettstein (2005) can be used to implement a MCST in SPNE with unique payoffs given by the Shapley value of the derived cooperative game (N, c), defined in (2.28), section 2.2.3.[13]

The original result in Pérez-Castrillo and Wettstein (2001) states that the Shapley value of any so-called *zero-monotonic* TU game (i.e., a game for which $v(S) + v(\{i\}) \leq v(S \cup \{i\})$ for any $S \subseteq N$ and $i \notin S$) can be implemented in SPNE using their bidding mechanism. Consequently, the Shapley value of

---

12. Ju and Wettstein (2009) present variations of the bidding mechanism where agents can re-enter the game. These variations implement the Shapley value, the equal surplus value (i.e., $\phi_i^{es} = v(\{i\}) + \frac{1}{n}(v(N) - \sum_{j \in N} v(\{j\})))$, and the consensus value (i.e., $\phi^c = 0.5\phi^{Sh} + 0.5\phi^{es}$).

13. See exercise 5.8.

all relevant derived games of the MCST model (being (N, c) defined in (2.28); $(N, \bar{c})$ defined in (2.29); and (N, c^*) defined in (2.30)) can be implemented by this mechanism, since these games are all zero monotonic.

Since the Folk solution, being the Shapley value of the game (N, c^*) derived from the irreducible cost matrix K^*, arguably satisfies a number of compelling fairness properties, Bergantiños and Vidal-Puga (2010) consider a version of the bidding mechanism that, loosely speaking, induces the game (N, c^*) and so supports implementation of the Folk solution defined in (2.32). This mechanism makes explicit use of the idea that agents connect sequentially to the root. Bids in the first stage of the mechanism are therefore not related to the willingness to pay for other agents being the proposer, but rather for connecting to the root.

The idea of the mechanism is as follows. In the first stage, agents are asked how much they are willing to pay every other agent if this agent connects to the root (source). A proposer is selected randomly among those with highest net offer. In the second stage, the proposer chooses a rooted spanning tree for some coalition $S \subseteq N$ (including herself) and proposes a cost allocation $y \in \mathbb{R}^S$. In the third stage, if the proposal is rejected, the proposer connects to the root, and every other agent pays her bid to the proposer (the proposer pays the cost of connecting to the root). All agents except for the proposer bargain again among themselves (but now with the option of connecting to the root via the proposer). If all agents in S accept the proposal, the proposed spanning tree is implemented, and remaining agents bargain among themselves, but now with the option of connecting to the root via agents in S.

The formal mechanism is defined recursively on the number n of agents: if $n = 1$, the agent i connects to the root and pays k_{0i}. Given that the rules of the game are known for at most $n - 1$ agents, we now define the bargaining game for the MCST problem with n agents, consisting of three stages:

1. Each agent $i \in N$ makes a bid $b^i_j \in \mathbb{R}$ for every agent $j \in N$ (what i is willing to pay j for connecting to the root). Agents bid simultaneously. For each $i \in N$, let $B_i = \sum_{j \neq i} b^j_i - \sum_{j \neq i} b^i_j$ be the net bid to agent i, and choose randomly a proposer $i^* \in \{\arg\max_i B_i\}$ maximizing the net bid.

2. The proposer i^* chooses a subset of agents $S_{i^*} \ni i^*$ and a rooted spanning tree $T \in \mathcal{T}^{S^0_{i^*}}$, and proposes payments $y_i \in \mathbb{R}^{S_{i^*}}$ covering the total cost of T.

3. Sequentially, the agents in $S_{i^*} \setminus \{i^*\}$ either accept or reject the proposal of i^*.

If some agent rejects the proposal, each agent $i \in N \setminus \{i^*\}$ pays $b^i_{i^*}$ to i^*, who connects to the root. Payoff to i^* is therefore $\sum_{i \neq i^*} b^i_{i^*} - c_{0i^*}$. Agents

$i \in N \setminus \{i^*\}$ continue to bargain, but now with the option of connecting to the root via i^* (payoff to these agents is therefore their payoff in the reduced game minus their payment to i^*).

If all agents accept the proposal, T is implemented, and the payoff to each $i \in S_{i^*}$ is $-y_i$. Agents in $N \setminus S_{i^*}$ continue to bargain, but now with the option of connecting to the root via agents in S_{i^*} (payoff to these agents equals their payoff in the reduced game).

In principle, no planner is needed to run the mechanism, but even if some trusted third party runs the procedure, there is no need for this third party to know connection costs in any of the stages.

Bergantiños and Vidal-Puga (2010) obtain the following result.

Theorem 5.7 At any SPNE, the MCST is implemented and payoffs are uniquely determined by the Folk solution ϕ^{Folk} defined in (2.32), i.e., the Shapley value of the cooperative game (N, c^*) based on the irreducible cost matrix.

Proof [Sketch of argument] Loosely speaking, the mechanism induces the game (N, c^*), since the lowest cost of connecting to the root for any coalition S, given that $N \setminus S$ is connected to the root already, is the irreducible cost $c^*(S)$. Thus, using a similar type of argument as in the proof of theorem 5.6, we can show that the following strategy constitutes a SPNE in the bidding game:

1. Each agent $i \in N$ bids $b^i_j = \phi_i^{Folk}(N, K) - \phi_i^{Folk}(N \setminus \{j\}, K^{+j})$ for every $j \neq i$.

2. The proposer i^* chooses the grand coalition N and a MCST of (N, K). Payments are determined by $y_i = b^i_{i^*} + \phi_i^{Folk}(N \setminus \{i^*\}, K^{+i^*})$ for all $i \in N \setminus \{i^*\}$.

3. Each agent in $S \setminus \{i^*\}$ accepts y if and only if $y_i \leq b^i_{i^*} + \phi_i^{Folk}(N \setminus \{i^*\}, K^{+i^*})$, where (S, K^{+T}) is the MCST problem obtained from (N, K), assuming agents in S have to connect to the root, and agents in $T \subseteq N \setminus S$ are connected to the root already (that is, $k^{+T}_{ij} = k_{ij}$ for all $i, j \in S$, and $k^{+T}_{i0} = \min_{j \in T^0} k_{ij}$, for all $i \in S$).

Clearly, equilibrium strategies lead to implementation of MCST. ∎

To gain some intuition, consider the following simple example.

Example 5.7 (Equilibrium strategies)   Two agents $\{1, 2\}$ want to connect to a root 0, with edge costs being $(k_{01}, k_{02}, k_{12}) = (10, 20, 2)$. The (unique) MCST is given by the graph $G^* = \{01, 12\}$ (shown below) with total cost 12.

If both agents play according to equilibrium strategies we get $b_2^1 = b_1^2 = 4$, implying that $B_1 = B_2 = 0$, and the proposer (randomly selected) will choose G^* and propose payments $y = (6, 6)$.

Now, one may ask why it is not optimal to reject any proposal since the agent who rejects can then connect to the root via the proposer at cost 2 (which is much lower than both standalone costs). Thus, a natural strategy seems to be to avoid being selected as proposer and then reject any proposal made to you.

However, here it is important to notice that the final payoff to the rejecting agent, say, agent 1, includes what needs to be paid to agent 2 for connecting to the root (i.e., his bid b_2^1). By selecting the proposer as the agent maximizing B_i, the mechanism ensures that the probability of being selected as the proposer is negatively correlated with the agent's willingness to pay the other agent for connecting to the root. So if an agent wants to reduce his probability of being selected as proposer, he has to bid higher (and thereby increase his final payment anyway): it is now clear that in equilibrium, we must have $B_1 = B_2 = 0$ with final payoffs equal to $(6, 6)$, as stated above.

Bergantiños and Lorenzo (2004) provide an empirical example of a decentralized network-formation process like the one used in the mechanism above, where agents connect sequentially to a source.

In their example, some Spanish villages were given the opportunity to connect individual households to local water reservoirs. The connection cost had to be covered by the villages, but the central authority was the network owner and was responsible for maintenance. Interestingly, the villages did not manage to implement an efficient network solution (here, a MCST). Instead, some front runners chose to connect to the source and among them, some actually managed to cooperate and share the cost. Subsequently, when it was clear to all that the system worked well, the remaining households decided to join the already existing network.

5.5 Strategyproofness

We now consider a situation where every user is characterized by an individual, and privately known, willingness to pay for network service. The system administrator is concerned with maximizing the total system welfare (i.e., aggregate willingness to pay minus the total cost of service). We suppose that the system administrator is able to set up a direct revelation mechanism, which elicits the network users' willingness to pay. In short, a direct revelation mechanism consists of an allocation rule (which determines who gets access to service) and a payment rule (which determines who pays what for the services obtained). The mechanism is based on the profile of reported willingness to pay from every user. Clearly, the administrator would like users to report their true willingness to pay, and in fact, to design the mechanism in such a way that truthful revelation is a dominant strategy for every user—called *strategyproofness*.

A classic result in mechanism design (Green et al., 1976) states that when users have quasilinear preferences, we cannot obtain strategyproofness, efficiency (here in the sense of welfare maximization), and budget balance of the payment rule simultaneously. In other words, insisting on strategyproofness implies a trade-off between efficiency and budget balance. Now, budget balance is in many ways a natural requirement in a decentralized scenario, since competition among service providers will drive prices down to actual costs. We therefore have to accept a certain loss in efficiency, but this loss should be as small as possible.

5.5.1 Moulin Mechanisms

Here we discuss a class of strategyproof mechanisms originating from Moulin (1999) and Moulin and Shenker (2001).

Let $N = \{1, \ldots, n\}$ be a finite set of agents. Every agent $i \in N$ has a willingness to pay for service $w_i \geq 0$, and agents are either served or not. The total cost of delivering service to any coalition of agents $S \subseteq N$ is given by a cost function $c : \mathcal{P}(N) \to \mathbb{R}_+$ (with $c(\emptyset) = 0$), which is assumed to be concave (see (1.6)).[14]

---

14. Recall that a concave cost function implies that the marginal cost of adding an agent to a given coalition of agents is decreasing with the size of that coalition.

From an overall system (or planner's) perspective, the question of who should get service depends on the aggregate welfare: some agents may simply be too costly to service compared to their willingness to pay. Aggregate welfare is maximized by servicing coalition S if and only if

$$\sum_{i \in S} w_i - c(S) \geq \sum_{i \in T} w_i - c(T),$$

for all $T \subseteq N$. In particular, for every coalition $T \subseteq N$, define the *standalone surplus* of T by

$$s(T, u) = \max_{K \subseteq T} \left\{ \sum_{i \in K} w_i - c(K) \right\}. \tag{5.9}$$

For every reported profile of willingness to pay $\omega \in \mathbb{R}^n_+$, a *direct mechanism* $\Lambda(\omega)$ produces a service vector $q = (q_1, \ldots, q_n)$, where $q_i \in \{0, 1\}$ (with $q_i = 1$ meaning that user i gets service, and $q_i = 0$ meaning that user i does not get service), and a payment vector $x = (x_1, \ldots, x_n)$ stating what every user i has to pay. Assume that agents have quasilinear utility: $w_i q_i - x_i$ for every $i \in N$.

In a decentralized setting, it is natural to impose some restrictions on a direct mechanism. Three normative requirements are as follows.

No subsidy   Cost shares are nonnegative: $x_i \geq 0$ for all $i \in N$.

Voluntary participation   Every agent reporting truthfully is guaranteed the status quo welfare level: no-service-no-payment ($q_i = x_i = 0$).

Consumer sovereignty   Every agent has a report ω_i, which regardless of the other agents' reports ω_{-i}, guarantees service to agent i ($q_i = 1$).

These weak requirements ensure that no agent is subsidized and can be worse off than not participating in the mechanism. They further guarantee that agents can always obtain service if they are willing to pay enough.

Furthermore, a direct mechanism is said to be *group strategyproof* if truthful reporting is a weakly dominating strategy for every group. Formally, a mechanism is group strategyproof if for every two profiles of reports w and w', and every coalition of agents $S \subseteq N$, satisfying $w_i = w'_i$ for all $i \notin S$, either (1) there is some $i \in S$, such that $w_i q'_i - x'_i < w_i q_i - x_i$, or (2) for all $i \in S$, it holds that $w_i q'_i - x'_i = w_i q_i - x_i$.

A direct mechanism is said to be *strategyproof* if it is group strategyproof for all singleton coalitions T, $|T| = 1$.

Let ξ be a budget-balanced cost-allocation rule that, for each cost function c and each coalition $S \subseteq N$, assigns an allocation of the total cost $c(S)$ to agents in S (i.e., $\sum_{i \in S} \xi_i(S) = c(S)$).

A *Moulin mechanism*, $M(\xi)$, induced by cost-allocation rule ξ is defined by the following stepwise procedure.

1. Collect reports, ω_i, for every agent in N.

2. Initialize $S = N$.

3. If $\omega_i \geq \xi_i(S)$, for every $i \in S$, then stop. The output is $q_i = 1$ for every $i \in S$, and $x_i = \xi_i(S)$ for every $i \in S$.

4. Let $i^* \in S$ be an agent with $\omega_{i^*} < \xi_{i^*}(S)$.

5. Set $S = S \setminus \{i^*\}$ and return to step 3. Set $q_{i^*} = x_{i^*} = 0$.

So a Moulin mechanism first tries to provide service to all agents, and if this turns out to be an individually rational solution (based on reported willingness to pay), everybody obtains service and pays according to the cost-sharing rule ξ. If some agent is assigned a cost share larger than the reported willingness to pay, this agent is excluded (does not get service and pays nothing), cost shares are recomputed for the set of remaining agents, and so forth, gradually diminishing the set of potential candidates who obtain service until individual rationality is achieved for everybody. Budget balance of ξ implies budget balance of $M(\xi)$, and when ξ is *population monotonic* (see section 2.2.4), agents for whom participation is too expensive at a given iteration of the mechanism will also find participation too expensive in later iterations since cost shares are decrease with population size. Therefore, it turns out that $M(\xi)$ will be strategyproof. In fact, Moulin (1999) proves the following result.

Theorem 5.8 For any population-monotonic cost-allocation rule ξ, the induced Moulin mechanism $M(\xi)$ is budget balanced; meets no subsidy, voluntary participation, and consumer sovereignty; and is group strategyproof.

Conversely, for any direct mechanism Λ, satisfying budget balance, no subsidy, voluntary participation, consumer sovereignty, and group strategyproofness, there exists a population-monotonic cost-allocation rule ξ such that the induced Moulin mechanism $M(\xi)$ is welfare equivalent to Λ.

As shown in chapter 2, many cost-allocation rules are population monotonic when the cost function is concave (for instance, the Shapley value and the Lorenz solution), so the mechanism designer has plenty of options when using Moulin mechanisms to obtain group strategyproofness.

However, the following example will show that Moulin mechanisms may produce highly inefficient outcomes.

Example 5.8 Recall the airport game from chapter 2, where every agent is characterized by a standalone cost C_i and the cost of any coalition $S \subseteq N$ is given by $c(S) = \max_{i \in S}\{C_i\}$. In particular, let $N = \{1, 2, 3\}$ and

$(C_1, C_2, C_3) = (1, 2, 3)$. In chapter 2 it was shown that c is concave. Let agents' willingness to pay be given by the profile $w = \left(\frac{5}{6}, \frac{5}{6}, \frac{10}{6}\right)$. Welfare maximum is obtained by letting all three agents be served at a total cost of 3 with surplus $s(N, w) = \frac{1}{3}$. Using a Moulin mechanism induced by the Shapley value, ξ^{Sh}, we obtain the following result.

- First iteration: Set $S = \{1, 2, 3\}$. We get $\xi^{Sh}(\{1, 2, 3\}) = \left(\frac{1}{3}, \frac{5}{6}, \frac{11}{6}\right)$, so for agent 3, we have $w_3 = \frac{10}{6} < \frac{11}{6} = \xi_3^{Sh}(\{1, 2, 3\})$. Therefore, set $q_3 = x_3 = 0$.
- Second iteration: Set $S = \{1, 2\}$. We get $\xi^{Sh}(\{1, 2\}) = \left(\frac{1}{2}, \frac{3}{2}\right)$, so for agent 2, we have $w_2 = \frac{5}{6} < \frac{3}{2} = \xi_2^{Sh}(\{1, 2\})$. Therefore, set $q_2 = x_2 = 0$.
- Third iteration: Set $S = \{1\}$. We get $\xi_1^{Sh}(\{1\}) = 1$, so for agent 1, we have $w_1 = \frac{5}{6} < 1 = \xi_1^{Sh}(\{1\})$. Therefore, set $q_1 = x_1 = 0$.

Clearly, the mechanism is performing highly inefficiently, since nobody obtains service, even though it is welfare optimizing to provide service to all three agents.

Given a cost-allocation rule ξ, the efficiency loss related to a given report profile ω can be measured as the difference between maximum welfare and the actual welfare achieved by the mechanism: $s(N, \omega) - \left(\sum_{i \in S(\xi, \omega)} w_i - c(S(\xi, \omega))\right)$, where $S(\xi, \omega) = \{i : q_i = 1 \text{ in } M(\xi)\}$ (i.e, the set of agents who obtain service by means of the mechanism). The worst-case efficiency loss is given by

$$L(\xi) = \sup_{\omega \in \mathbb{R}_+^n} \left[s(N, \omega) - \left(\sum_{i \in S(\xi, \omega)} w_i - c(S(\xi, \omega)) \right) \right], \tag{5.10}$$

and a natural question is therefore whether we can find a cost-allocation rule that minimizes the worst-case efficiency loss $L(\xi)$. Moulin and Shenker (2001) provide an answer.

Theorem 5.9   In the class of population-monotonic cost-allocation rules, the Shapley value, ξ^{Sh}, has the unique minimal worst-case efficiency loss, with potential

$$L(\xi^{Sh}) = \left(\sum_{S \subseteq N} \frac{(|S| - 1)!(n - |S|)!}{n!} c(S) \right) - c(N) \tag{5.11}$$

(recall that $0! = 1$).

Example 5.8 (continued)   Recall example 5.8 above, where the efficiency loss is $\frac{1}{3}$. In this example it is simple to see that other profiles w would have generated even larger losses. Clearly the profile $w = (1 - \varepsilon, 1.5 - \varepsilon, 1.83 - \varepsilon)$

will generate the maximal loss, since no agent will obtain service, but the potential welfare is $1.33 - 3\varepsilon$. Using (5.11) to determine $L(\xi^{Sh})$, we get exactly $\frac{6}{3}$ (summing over singleton coalitions) plus $\frac{8}{6}$ (summing over 2-coalitions) plus 1 (for the grand coalition) minus the total cost of 3, yielding 1.33.

5.5.2 Engineering Applications

Moulin-type mechanisms have potential for application in a number of different areas related to network control and administration. In this section I briefly sketch a couple of these applications concerning multicast cost sharing (Feigenbaum et al., 2001; Archer et al., 2004) and bandwidth on demand (Shi et al., 2015).

In multicast transmission, a multicast flow emanates from a source node 0 to a set of receivers N located at different nodes in a multicast tree $T(N)$, where $T(N)$, can be considered as the smallest tree required to reach agents in N. Transmitting over each link l in the tree has an associated cost $c_l \geq 0$, which is known by all receivers as well as by the network administrator, and each receiver $i \in N$ assigns a value w_i to receiving the transmission. The value w_i is private information of receiver i. The cost function associated with transmission to any coalition of receivers, $S \subseteq N$, is concave, as shown in chapter 2.

In many cases it seems natural to assume that the system administrator is concerned with overall system welfare, in particular, if the network is regulated. Therefore, Moulin mechanisms can be used by the system administrator to determine who will receive transmissions and how much they are going to pay. For example, using the Shapley value as a cost-sharing rule produces budget balance and group strategyproofness: that is, no individual agent or group of agents can game the system by misreporting their true value of receiving transmissions. Moreover, using the Shapley value minimizes the worst-case efficiency loss as noted above.

A Moulin-type mechanism associated with a Shapley cost-sharing rule has also been considered in connection with bandwidth on demand. A cloud provider operates a network spanning multiple data centers. Links are owned by ISPs and leased to the cloud provider. Links have limited bandwidth capacity. Users arrive at certain times t_i with data transfer requests of size B_i between some pair of data centers, within d_i time slots. If the transfer is finished in the specified time frame, users have willingness to pay w_i. Users' willingness to pay w_i is private information.

The cloud provider wants to find a mechanism that works like an iterative ascending auction. Given users' "bids" ω_i, the mechanism accepts or rejects user i's transfer request ($q_i \in \{0, 1\}$), determines how to route the data traffic (B_i, d_i), and how much the user has to pay (x_i). The cloud provider targets maximization of social welfare (here the sum of the cloud provider's revenue and the total user utility) under technical feasibility constraints due to limited capacity. Users have quasilinear utility. Traffic is optimized to minimize the cost paid to ISPs, and this cost is subsequently allocated to accepted users.

However, under standard ISP charging models (like a 95th percentile or max-traffic charge), the cost function associated with coalitions of users, $c(S)$, is not concave, so straightforward application of Moulin mechanisms is prevented. But under various assumptions on users' willingness to pay, Shi et al. (2015) show that defining cost shares x_i according to the Shapley value (here defined as the average marginal ISP charge incurred by user i's traffic) induces a Moulin-like mechanism that is truthful in bidding, budget balanced in expectation, individually rational, and limited in efficiency loss, for on-line traffic scheduling.

5.6   Exercises

5.1. (Roughgarden, 2007) Consider the following (Pigou) example. Two parallel roads (upper and lower) leads from source s to sink t, where $c(x)$ is the travel time (cost) of the road (arc) when a fraction x of the agents is using it. So the upper road has a fixed travel time of, say, 1 hour, while travel time on the lower road depends on the traffic.

$c(x) = 1$

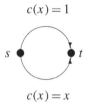

$c(x) = x$

What is the unique Nash equilibrium in this case? How would a social planner route the traffic? What are the resulting PoA and PoS?

Now, what happens if the linear cost $c(x) = x$ is replaced by a nonlinear cost $c(x) = x^p$: what will be the optimal flow in this case? Find PoA when p tends to infinity.

5.2. (Anshelevich et al., 2008) Recall theorem 5.1. Show that this theorem still holds if the constant cost of each edge, c_e, is replaced by a nondecreasing concave cost function $c_e(x)$, where x is the number of agents using the edge. (Hint: Use the same line of argument as in the sketch of the proof for theorem 5.1).

5.3. (Roughgarden and Schrijvers, 2016) Consider a network cost-sharing game, but in a situation where the agents differ, say, in terms of edge usage.

Let $w_i > 0$ be the usage of agent $i \in M$ and $w \in \mathbb{R}^n$ the profile of agents' usage "weights." Edge costs are a function of total usage: in particular, for every edge e, let $c_e(T) = \alpha \sum_{i \in T} w_i$, for $\alpha > 0$, be the cost when coalition $T \subseteq M$ uses e. For every edge e and set of its users S_e, the network designer uses the Shapley value of the cost game (S_e, c_e) as cost-sharing protocol (i.e., the cost of every edge is shared according to a weighted average of marginal costs for every agent, as in the usual (unweighted) Shapley value: notice that this is not the same as the protocol (5.6)). Illustrate the use of the Shapley value as a protocol in this case. Does the associated network cost-sharing game always have at least one pure-strategy Nash equilibrium? Find the worst-case PoA.

5.4. (Epstein et al., 2009) Consider the network cost-sharing game with a directed graph and Shapley protocol (5.6). Prove that if all agents want to connect the same source-sink pair, the game possesses a strong Nash equilibrium.

Moreover, consider a variation of the graph in example 5.3, where the last two agents, $m - 1$ and m, have a joint path with cost $2/m$ from s to their respective sinks t_{m-1} and t_m. Show that in this game (using the Shapley protocol), we have strong PoS > PoS.

5.5. Consider the Jackson-Wolinsky model, and recall example 5.5. Show that the efficient network

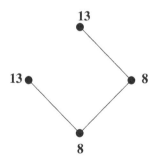

is pairwise stable using the extended Myerson value (4.29) as the allocation rule.

5.6. A value function v is monotonic if $v(G') \leq v(G)$ when $G' \subseteq G$. Does theorem 5.3 hold for the restricted class of monotonic value functions?

5.7. (Dutta and Mutuswami, 1997) Try to prove the following result: *There is no anonymous and component-balanced allocation rule ϕ such that for each value function v, at least one efficient network is weakly stable (i.e., the network results from a coalition-proof Nash equilibrium of the game (5.7).* (Hint: Base your argument on the three-agent case, where the value function is given by $v(G^N) = v(\{ij\} = 1$, and $v(\{ij, jk\}) = 1 + 2\varepsilon$, where $0 < \varepsilon < 1/12)$. Discuss the relation, of this result to theorem 5.4.

5.8. (Bergantiños and Vidal-Puga, 2010) Show how the bidding mechanism of Pérez-Castrillo and Wettstein (2005) can be used to implement a MCST in SPNE, with unique payoffs given by the Shapley value of the derived cooperative game (N, c) defined in (2.28).

References

Acemoglu, D., and A. Ozdaglar (2007), "Competition and Efficiency in Congested Markets," *Mathematics of Operations Research* 32: 1–31.

Altman, E., T. Boulogne, R. El-Azouzi, T. Jimenez, and L. Wynter (2006), "A Survey on Networking Games in Telecommunication," *Computers and Operations Research* 33: 286–311.

Anshelevich, E., A. Dasgupta, J. Kleinberg, E. Tardos, T. Wexler, and T. Roughgarden (2008), "The Price of Stability for Network Design with Fair Cost Allocation," *SIAM Journal of Computing* 38: 1602–1623.

Archer, A., J. Feigenbaum, A. Krishnamurthy, R. Sami, and S. Shenker (2004), "Approximation and Collusion in Multicast Cost Sharing," *Games and Economic Behavior* 47: 36–71.

Bala, V., and S. Goyal (2000), "A Noncooperative Model of Network Formation," *Econometrica* 68: 1181–1229.

Bergantiños, G., and L. Lorenzo (2004), "A Non-cooperative Approach to the Cost Spanning Tree Problem," *Mathematical Methods of Operations Research* 59: 393–403.

Bergantiños, G., and J. Vidal-Puga (2010), "Realizing Fair Outcomes in Minimum Cost Spanning Tree Problems through Non-cooperative Mechanisms," *European Journal of Operational Research* 201: 811–820.

Bilò, V., M. Flammini, and L. Moscardelli (2014), "The Price of Stability for Undirected Broadcast Network Design with Fair Cost Allocation Is Constant," *Games and Economic Behavior*, in press.

Chen, H.-L. T. Roughgarden, and G. Valiant (2010), "Designing Network Protocol for Good Equilibria," *SIAM Journal of Computing* 39: 1799–1832.

DaSilva, L. A. (2000), "Pricing for QoS-Enabled Networks: A Survey," *IEEE Communications Surveys & Tutorials* 3.

Dutta, B., and M. O. Jackson (2003), "The Stability and Efficiency of Directed Communication Networks," in, B. Dutta et al., (eds), *Networks and Groups*. New York: Springer.

Dutta, B., and S. Mutuswami (1997), "Stable Networks," *Journal of Economic Theory* 76: 322–344.

Epstein A., M. Feldman, and Y. Mansour (2009), "Strong Equilibrium in Cost Sharing Connection Games," *Games and Economic Behavior* 67: 51–68.

Feigenbaum, J., C. Papadimitriou, and S. Shenker (2001), "Sharing the Cost of Multicast Transmissions," *Journal of Computer and System Sciences* 63: 21–41.

Gopalakrishnan, R., J. R. Marden, and A. Wierman (2014), "Potential Games Are Necessary to Ensure Pure Strategy Nash Equilibria in Cost Sharing Games," *Mathematics of Operations Research* 39: 1252–1296.

Green, J., E. Kohlberg, and J. J. Laffont (1976), "Partial Equilibrium Approach to the Free Rider Problem, *Journal of Public Economics* 6: 375–394.

Hart, S., and A. Mas-Colell (1996), "Bargaining and Value," *Econometrica* 64: 357–380.

Holzman, R., and N. Law-yone (1997), "Strong Equilibrium in Congestion Games," *Games and Economic Behavior* 21: 85–101.

Holzman, R., and N. Law-yone (2003), "Network Structure and Strong Equilibrium in Route Selection Games," *Mathematical Social Sciences* 46: 193–205.

Jackson, M. O. (2003), "The Stability and Efficiency of Economic and Social Networks," in Koray, S., and M. R. Sertel (eds.), *Advances in Economic Design*. New York: Springer.

Jackson, M. O. (2005), "Allocation Rules for Network Games," *Games and Economic Behavior* 51: 128–154.

Jackson, M. O. (2008), *Social and Economic Networks*. Princeton, NJ: Princeton University Press.

Jackson, M. O., and A. Watts (2002), "The Evolution of Social and Economic Networks," *Journal of Economic Theory* 106: 265–295.

Jackson, M. O., and A. Wolinsky (1996), "A Strategic Model of Social and Economic Networks," *Journal of Economic Theory* 71: 44–74.

Ju, Y., and D. Wettstein (2009), "Implementing Cooperative Solution Concepts: A Generalized Bidding Approach," *Economic Theory* 39: 307–330.

Koutsoupias, K., and C. Papadimitriou (1999), "Worst-Case Equilibria," in *Proceedings of the 16th Annual Symposium on Theoretical Aspects of Computer Science*, 404–413, vol. 1563, *Lecture Notes in Computer Science*. Berlin: Springer.

Monderer, D., and L. S. Shapley (1996), "Potential Games," *Games and Economic Behavior* 14: 124–143.

Moulin, H. (1999), "Incremental Cost Sharing: Characterization by Coalition Strategy-proofness, *Social Choice and Welfare* 16: 279–320.

Moulin, H., and S. Shenker (2001), "Strategy-Proof Sharing of Submodular Costs: Budget Balance versus Efficiency," *Economic Theory* 18: 511–533.

Mutuswami, S., D. Pérez-Castrillo, and D. Wettstein (2004), "Bidding for the Surplus: Realizing Efficient Outcomes in Economic Environments," *Games and Economic Behavior* 48: 111–123.

Ozdaglar, A., and R. Srikant (2007), "Incentives and Pricing in Communication Networks," in Nisan et al. (eds.), *Algorithmic Game Theory*. Cambridge: Cambridge University Press.

Papadimitriou, C. (2001), "Algorithms, Games, and the Internet," *Proceedings of the 33rd Annual ACM Symposium on the Theory of Computing*, 749–753. New York: ACM Digital Library.

Pérez-Castrillo, D., and D. Wettstein (2001), "Bidding for Surplus: A Non-cooperative Approach to the Shapley Value," *Journal of Economic Theory* 100: 274–294.

Pérez-Castrillo, D., and D. Wettstein (2005), "Forming Efficient Networks," *Economics Letters* 87: 83–87.

Rosenthal, R. W. (1973), "A Class of Games Possessing Pure-Strategy Nash Equilibria," *International Journal of Game Theory* 2: 65–67.

Roughgarden, T. (2007), "Routing Games," in Nisan et al. (eds), *Algorithmic Game Theory*. Cambridge: Cambridge University Press.

Roughgarden, T. and O. Schrijvers (2016), Network cost-sharing without anonymity, *ACM Transactions on Economics and Computation*, 4.

Serrano, R. (2008), "Nash Program," in S. N. Durlauf and L. Blume (eds.), *The New Palgrave Dictionary of Economics*, (2nd ed.), Basingstoke, UK: Mcmillan.

Shi, W., C. Wu, and Z. Li (2015), "A Shapley-Value Mechanism for Bandwidth on Demand between Datacenters," *IEEE Transactions on Cloud Computing*, to appear.

Slikker, M. (2007), "Bidding for Surplus in Network Allocation Problems," *Journal of Economic Theory* 137: 493–511.

van den Brink, R., G. van der Laan, and N. Moes (2013), "A Strategic Implementation of the Average Tree Solution for Cycle-Free Graph Games," *Journal of Economic Theory* 148: 2737–2748.

6 Efficient Implementation

We now study a scenario where a benevolent social planner wants to implement an efficient network centrally, but information is asymmetric: agents are fully informed; the planner is not. The role of the planner is to design an institution (a so-called *game form*, or *mechanism*) through which agents' strategic interactions will lead to a desired outcome—in our case, typically an efficient network and a compelling allocation of its cost or benefit. This is no easy task if we want full implementation, in the sense that for every possible state, all equilibrium outcomes are desired outcomes. General requirements for full implementability are found in, for example, Maskin and Sjöström (2002) and Jackson (2001), both providing a survey of the literature on implementation theory.

We therefore start out being less ambitious and search for game forms that implement an efficient network by truthful reporting; that is, truthful revelation of costs and benefits constitutes a Nash equilibrium of the induced game. This does not mean that lying cannot be an equilibrium strategy that leads to an efficient network, nor does it exclude the possibility that other equilibria may lead to inefficient networks.

In case of the MCST model, we show that a particular game form implements an efficient network by truthful reporting provided that the cost-allocation rule is reductionist and cost monotonic. From chapter 2 we know that there exists a large class of rules satisfying these two properties, including use of the Folk solution (2.32), so the planner has several options when designing the rules of the game. Moreover, all Nash equilibria of the suggested game form actually lead to implementation of MCSTs (i.e., worst-case PoA is 1) when a certain weak condition is satisfied.

The options for efficient implementation are more limited when generalizing the model to arbitrary connection demands. Using a straightforward adaptation of the same game form, relevant cost-allocation rules now violate even individual rationality, in the sense that network users may end up paying more than their standalone costs. Implementing the efficient network by truthful reporting is only possible if the planner uses an allocation rule that ensures individual cost shares only depend on the total network cost and the other agents' connection demands. So ensuring network efficiency by truthful reporting now implies compromising with distributional fairness in the form of standalone upper bounds.

If the planner has full information about connection demands, the worst-case PoA becomes 1. However, when the planner does not know connection demands, even full knowledge of connection costs does not prevent Nash equilibrium networks from perhaps being highly inefficient, so PoA can be unbounded in this case. Looking at implementation in a decentralized scenario (à la the network cost-sharing games in chapter 5), somewhat similar results can be obtained.

Since the connection network model implicitly assumes that network users' willingness to pay for connectivity is unlimited, it is natural to examine whether a limited willingness to pay will influence the possibilities of implementing efficient networks. For limited willingness to pay, the planner aims to implement a welfare-maximizing network that may exclude certain users if their willingness to pay is too low. We show that desirable outcomes, in the form of a welfare-maximizing network and an allocation of its cost respecting the limited willingness to pay of all agents, are Nash implementable in the sense that there exists a game form for which an outcome is a Nash equilibrium if and only if it is desirable. Such a game form can be constructed along the lines of the canonical game form in Maskin (1999). However, no selection from the correspondence of desirable outcomes is Nash implementable, so the planner cannot use a specific allocation rule but instead has to be flexible when allocating cost case by case (albeit respecting agents' limited willingness to pay every time). Moreover, the cost allocation of a desirable outcome may be quite unfair with respect to its derived allocation of obtained welfare. We can show that when the total welfare is positive, no Nash implementable solution can guarantee a positive share of this welfare to all agents.

If we are willing to accept game forms building on sequential reporting by agents, it turns out to be possible to implement a selection from the correspondence of desirable outcomes in SPNE. Although multiple such equilibria may exist, the expected payoff is uniquely determined by the Shapley value.

Throughout this chapter we focus on implementation in a complete information setting. Arguably, this is a somewhat strong assumption. In practice, agents typically have only partial information about cost, or know their own connection demand and willingness to pay, as well as maybe that of neighboring agents. While there exist general results concerning implementability in a

Bayesian framework, specific analysis of network implementation has been largely ignored. In general the literature on implementation in networks is much less developed than the axiomatic literature discussed in the previous chapters, making the former a potentially fruitful topic for future contributions.

6.1   Truthful Reporting: Preliminary Examples

When information is asymmetric, the uninformed planner relies on information reported by the agents. A key question is under which circumstances we can expect agents to report the information truthfully to the planner.

We start out by sketching strategic behavior by informed agents in two of the classic models from the previous chapters: the MCST model and the closely related capacity networks model. Specifically, we assume that the agents know all connection costs, and the planner does not. Moreover, both agents and planner know all agents' connection demands.

6.1.1   The MCST Model

Recall the MCST model from section 2.2, where a set of agents wants connection to a single source (denoted 0). The MCST model is a special case of the minimum-cost connection networks model from section 4.4, where agents can demand connectivity between any two nodes in the network. Assuming that all agents know all connection costs (i.e., the full cost matrix), while the planner does not, introduces the possibility of strategic manipulation by some agents. Since the planner wants to implement an efficient network centrally, she relies on cost information reported by the agents. Yet it is clear that agents may have opposing interests in reporting the true costs, depending on how the planner intends to charge the agents based on their reports (i.e., depending on the planner's choice of cost-allocation rule).

A simple example includes two agents, Ann and Bob, sharing network costs using the Bird rule (2.17). Suppose true costs are given by $K = (k_{0A}, k_{0B}, k_{AB}) = (2, 3, 1)$. The unique efficient network is thus $G = \{0A, AB\}$, with a total cost of 3, as illustrated below.

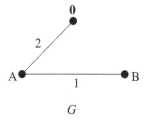

G

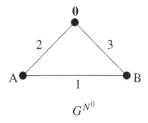

G^{N^0}

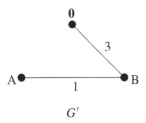

G'

So using the Bird rule, Ann pays 2, and Bob pays 1. Ann and Bob know all connection costs, but the planner does not. Suppose both Ann and Bob report all costs to the planner, and Bob is always reporting truthfully. The planner uses agents' cost reports to select a cost-minimizing network and always allocates the total cost of the implemented network on the basis of the cost reports. Now, if Ann's report can influence the planner's network choice (for instance, if the planner estimates connection costs as the mean of reported costs), she therefore has incentive to claim that her direct source connection $0A$ is more expensive than 2. If she can make the planner estimate a cost higher than 3 for her source connection $0A$, the planner will choose the (inefficient) network $G' = \{0B, AB\}$, since G' appears to be a MCST based on the reported costs. If Alice succeeds, she will therefore only pay 1, despite the increase in total network cost.

Suppose the planner obtains information about the true connection cost of the selected network when constructing it. Then it appears that the planner cannot observe that Ann has been lying. Indeed, implementing the inefficient network G' only reveals the true part of the reported costs from both agents. So punishing agents for lying will not make Ann report truthfully in this case.

It is clear that we may run into similar problems using any kind of network-dependent allocation rule (like the Bird rule), since such rules induce agents to have preferences about network configurations, and this will, in turn, make them try to manipulate the planner's network choice through their cost reports.

To see why manipulation by misreporting costs may also occur for rules that do not depend on the realized network structure, consider the more subtle rule: allocate costs in proportion to the standalone costs related to the irreducible cost matrix of the problem. Specifically, use cost shares $x_i = \frac{k^*_{0i}}{\sum_j k^*_{0j}} v(N, K)$, for every agent i, where $v(N, K)$ is the total cost of a MCST $T^{\min}(N, K)$, and K^* is the irreducible cost matrix associated with the problem (N, K).

For instance, take the following three-agent problem, where costs are given in the complete graph below.

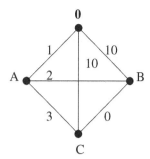

Clearly, the unique MCST is $G = (0A, AB, BC)$ with a total (minimum) cost of 3. The corresponding irreducible costs are given by the following figure.

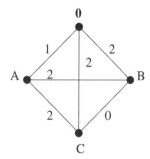

Thus, Ann pays $\frac{1}{5} \times 3 = \frac{3}{5}$, while Bob and Carl both pay $\frac{2}{5} \times 3 = \frac{6}{5}$, if the planner implements the efficient network G.

Now suppose that Bob and Carl report the truth, and the planner estimates the connection costs as the average of the reported cost matrices. Then Ann can gain by reporting a cost larger than 5 for the edge AB and the true connection costs for the other edges. Indeed, the planner now estimates the cost of edge AB to be larger than 3, and hence selects the spanning tree $G' = (0A, AC, CB)$ as the unique MCST of the *estimated* cost matrix, $K^e = (1, 10, 10, k^e_{AB}, 3, 0)$, where $k^e_{ab} > 3$. The associated (estimated) irreducible cost matrix thus becomes $K^{e*} = (1, 3, 3, 3, 3, 0)$, so Ann now pays $\frac{1}{7} \times 4 = \frac{4}{7} < \frac{3}{5}$ and thus gains from misreporting.

The example indicates that we need allocation rules to be monotonic in connection costs (at least in the sense that the payment of agent i is weakly increasing in i's connection costs k_{ij}, for all j) for there to be any hope of inducing the agents to report truthfully.

6.1.2 Capacity Networks

The capacity network model (from section 4.3) has many similarities with the MCST and MCCN models. Assume that capacities k_{ij} are known by the agents but not by the planner. As above, suppose the planner uses a mechanism in which each pair of agents reports their capacity need (simultaneously and independently), and the planner then estimates the capacity matrix K^e based on the reported capacities (e.g., as the average), selects a minimum-capacity network, and allocates the resulting cost using a preannounced cost-allocation rule.

Monotonicity of the allocation rule will prevent agents from reporting capacities beyond their true need, simply because it makes them pay more than what is necessary. However, it does not prevent agents from underreporting. Repeating an example provided in Bogomolnaia et al. (2010), let us consider a case with four agents $\{1, 2, 3, 4\}$ and (true) capacity needs, $k_{ij} = 10$, for all $ij \neq 12$, and $0 < k_{12} < 10$. It is clear that a minimum-capacity network will never include the edge e_{12} but will consist of a three-edge spanning tree with a total cost of 30 and a capacity of 10 for each edge. Therefore agents 1 and 2 can actually gain by underreporting their capacity needs for edge e_{12}. By reporting $\sigma_{12} < k_{12}$, this will have no influence on the chosen minimum-capacity network, so their demands will still be met, and since the cost-allocation rule is monotonic in capacity, they might now pay less.

However, if we further require that the monotonic rule should be based on data from the irreducible capacity matrix, it is possible to eliminate gains from underreporting as well. Indeed, in general $k_{ij} \leq k_{ij}^*$ for any pair of agents i and j. Since $k_{12} \leq k_{12}^*$, and agents 1 and 2 report $\sigma_{12} < k_{12}$, underreporting has no influence on the cost shares, because these are determined by K^*. In general, if $k_{ij} = k_{ij}^*$ and $\sigma_{ij}^* < k_{ij}^*$, agents i and j will not have their true capacity need covered.

In the next section we will clarify the connection between cost-allocation rules based on the irreducible cost matrix and the property of monotonicity, using a specific game form for the MCST model.

6.2 Implementation: The MCST Model

Following Hougaard and Tvede (2012) we revisit the MCST model in a situation where the planner (who is unaware of connection costs) wants to implement an efficient network, while the agents (who know the cost matrix) want to satisfy their connectivity demand with the lowest possible payment. It seems natural to assume that everybody, including the planner, knows that all agents want connection to the root (source) as well as the location of every agent in the network. So agents' connection demands can be considered public information.

Recall from chapters 1 and 2 that an allocation problem consists of a pair (N, K), where N is a set of agents, and K is the associated edge-cost matrix of the complete graph with $N^0 = N \cup \{0\}$ nodes, with 0 being the root. Moreover, let $\mathcal{T}$ be the set of spanning trees and $\mathcal{T}^{min}(K)$ be the set of MCSTs for a given problem.

Let Λ denote set of allocation problems and their spanning trees, that is, $(N, K, T) \in \Lambda$ if and only if (N, K) is an allocation problem and $T \in \mathcal{T}$ is a spanning tree for (N, K).

6.2.1 The Game Form

To implement the efficient network, the planner uses the following four-stage mechanism, Φ:

1. The planner announces the rules of the game: an estimation rule and a cost-allocation rule.

2. Every agent reports, simultaneously and independently, all connection costs (i.e., a cost matrix).

3. Based on the reported cost matrices from all agents, the planner uses the estimation rule to produce an estimated cost matrix and randomly selects a MCST (efficient network), given the estimated cost matrix.

4. The planner implements the selected network and allocates the observed true costs using the announced cost-allocation rule.

Agents may misreport costs, which influences the planner's choice of network, but once a network is chosen for implementation, the planner will learn (and distribute) the true connection costs of that spanning tree. Think of this as the planner learning the true costs of constructing links while establishing the network.

Therefore it is natural for the planner to use either an allocation rule that is related directly to the structure of the implemented network, or a *reductionist* allocation rule based on the irreducible cost matrix, because the planner, knowing the true costs of the implemented spanning tree, can use the construction of the irreducible cost matrix to reverse engineer the remaining connection costs. That is, for a given $(N, K, T) \in \Lambda$, $K^*(T)$ represents the smallest cost matrix for which the implemented network T will be compatible with an efficient (cost-minimizing) network.

6.2.2 The Setup

Formally, let $\phi : \Lambda \to \mathbb{R}^n$ be the announced budget-balanced allocation rule, which is further assumed to be continuous in K, and reductionist in the sense that

$$\phi_i(N, K, T) = \phi_i(N, K^*(T), T) \tag{6.1}$$

for all $T \in \mathcal{T}$.

Moreover, let $\tau : \mathbb{R}^n \to \mathbb{R}$ be the announced estimation rule used to estimate the cost of each element in the cost matrix. Assume that the estimation rule τ takes values between the reported minimum and maximum costs:

$$\tau(\sigma_{jl}^1, \ldots, \sigma_{jl}^n) \in [\min\{\sigma_{jl}^1, \ldots, \sigma_{jl}^n\}, \max\{\sigma_{jl}^1, \ldots, \sigma_{jl}^n\}].$$

Moreover, assume that the estimation rule is *upward unbounded*, that is, $\lim_{\sigma_{jl}^i \to \infty} \tau(\sigma_{jl}^1, \ldots, \sigma_{jl}^n) = \infty$ for all $i \in N$. Note that the latter assumption implies that τ cannot be defined as taking the minimum of the reported costs. A cost estimate based on the maximum is an example of an upward-unbounded estimation rule.

Let $\sigma \in (\mathbb{R}^{N^0(2)})^n$, where $\sigma = (\sigma^1, \ldots, \sigma^n)$ and $\sigma^i = (\sigma_{01}^i, \ldots, \sigma_{(n-1)n}^i)$ for all $i \in N$, be a profile of individually reported cost matrices. For a given profile of cost reports σ, denote by $K^e(\sigma)$ the estimated cost matrix with elements $k_{jl}^e = \tau(\sigma_{jl}^1, \ldots, \sigma_{jl}^n)$ for all j, l.

Agents choose cost reports strategically, minimizing their expected payments given the rules of the game (ϕ, τ). That is, every agent $i \in N$ chooses σ^i to minimize

$$\sum_{T \in \mathcal{T}^{min}(K^e(\sigma))} \frac{\phi_i(N, K^*(T), T)}{|\mathcal{T}^{min}(K^e(\sigma))|}, \tag{6.2}$$

since all MCSTs (given the estimated cost matrix $K^e(\sigma)$) are equally likely to be implemented by the planner.

Example 6.1 (Illustrating the setup) Recall the problem considered in section 6.1.1, where costs are given by $K = (k_{0A}, k_{0B}, k_{AB}) = (2, 3, 1)$ with $G = \{0A, AB\}$ as the unique cost-minimizing network, as shown below.

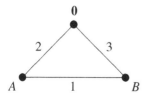

This time, assume the planner uses the Folk solution (see (2.32)) as the cost-allocation rule ϕ, and the arithmetic mean as the (upward unbounded) estimation rule τ. Suppose that Ann reports $\sigma^A = (4, 1, 1)$, while Bob reports the truth: $\sigma^B = (2, 3, 1)$. The planner then estimates the cost matrix $K^e = (3, 2, 1)$, and selects what is believed to be the MCST $\overline{G} = \{0B, AB\}$. Upon learning the true cost of $\overline{G}$ (i.e, $k_{0B} = 3$, and $k_{AB} = 1$), the planner finds the associated irreducible cost matrix $K^{*e} = (3, 3, 1)$ (as if $\overline{G}$ were the true MCST). Consequently, the network cost of 4, is shared as $(2, 2)$ using the Folk solution. Clearly, this is not a Nash equilibrium, since Ann, by reporting the true cost, can reduce her payment to 1.5 (because the true irreducible cost matrix is $K^* = (2, 2, 1)$ with Folk solution $(1.5, 1.5)$). Also, given Ann's report, Bob can obtain the same reduction in payment by reporting $\sigma'^B = (0, 1, 5)$, and thereby make sure that the estimated cost matrix becomes the true cost matrix. Both (σ^A, σ'^B) and truthful reporting by both agents are examples of Nash equilibria leading to implementation of the efficient network.

6.2.3 Implementation

From the above discussion it is evident that to prevent manipulation, the cost-allocation rule has to be monotonic in connection costs. Here, we relate monotonicity directly to the irreducible cost matrix K^*, since allocation rules are reductionist by definition.

Monotonicity For two spanning trees T and T' where $K^*(T) \leq K^*(T')$, we have $\phi(N, K^*(T), T) \leq \phi(N, K^*(T'), T')$.

Recall from chapter 1 that the irreducible cost matrix is minimal for a MCST compared to any other spanning tree. Thus, if a reductionist cost-allocation rule satisfies monotonicity, no agent (or coalition of agents) has incentive to deviate from truthful reporting. In fact, truthful reporting is a (strong) Nash equilibrium only if the reductionist cost-allocation rule is monotonic. Indeed, assume that the reductionist rule is not monotonic; then some agent (or coalition of agents) will benefit from implementation of an inefficient network. This agent (coalition) can always make the planner choose such a network by lying about the costs of all other connections than those of the particular spanning tree he wants (since the estimation rule is upward unbounded). We record this result of Hougaard and Tvede (2012) in the following theorem.[1]

Theorem 6.1 Truthful reporting of costs is a (strong) Nash equilibrium for every allocation problem if and only if the allocation rule ϕ is monotonic.

---

1. Hougaard and Tvede (2012) also extend theorem 6.1 to a Bayesian setting.

Remark   Note that since the allocation rule is reductionist, the monotonicity notion can actually be weakened to standard cost monotonicity (see section 2.2.4), that is, for all pairs of cost matrices K and K', where $k_{ij} > k'_{ij}$ for $i, j \in N^0$, $k'_{nm} = k_{nm}$ for all $\{n, m\} \neq \{i, j\}$, and $\phi_z(N, K) \geq \phi_z(N, K')$, for all $z \in N \cap \{i, j\}$. Indeed, if a spanning tree different from a MCST is chosen by the planner, all agents i will have a link cost in the associated irreducible matrix that is strictly increasing for some j. Thus, for reductionist rules, cost monotonicity implies monotonicity in the strong form defined above.

Recall from chapter 2 that a large class of reductionist rules satisfies monotonicity, given by the class of obligation rules (including the Folk solution, i.e., the Shapley value of the irreducible game (2.32)). So the planner has several options when designing the specific rules of the mechanism Φ.

However, truthful reporting is not a dominant strategy, as illustrated in example 6.1 above. In fact, by strategic misreporting, any agent can always ensure that the efficient network is implemented, given the reports of the other agents. In this sense, false reports actually play a positive role in the present setup. Together with the fact that the planner cannot always observe whether agents have been lying, it demonstrates that it is pointless for the planner to attempt to punish misreporting. Moreover, an important consequence is that *any Nash equilibrium will lead to implementation of the true MCST*. That is, if σ is a Nash equilibrium, and $T \in \mathcal{T}^{min}(N, K^e(\sigma))$, then $T \in \mathcal{T}^{min}(N, K)$. To see this, suppose that a MCST for a profile of Nash equilibrium reports $\bar{\sigma}$ is not a true MCST. Then the total cost is higher than the minimal cost, and hence some agent will pay more than if the true MCST were implemented (by budget balance). Now, this agent may change his strategy such that by lying about all other connection costs (other than those of the true MCST), he can make the planner choose the true MCST (again this is possible since τ is upward unbounded), which contradicts the notion that the reported costs $\bar{\sigma}$ constitute a Nash equilibrium.

Recall the definition of the price of anarchy (PoA) from chapter 5. Since any Nash equilibrium will lead to implementation of the true MCST, worst-case PoA equals 1 using the four-stage mechanism Φ, with τ being upward unbounded.

6.3   Implementation: The MCCN model

Using the same four-stage mechanism Φ, Hougaard and Tvede (2015) extend the analysis of efficient implementation to the more general minimum-cost connection network (MCCN) model. Here, agents have connection demands in the form of some pair of nodes they want connected, either directly or indirectly (see section 4.4). However, since individual connection demands are no longer public information, the agents must also report their connection demand in addition to the connection costs, in stage 2 of the mechanism. In the MCCN model, agents therefore have the additional option of reporting their connection demand strategically. As before, agents want to satisfy their connection demand with minimum payment, and the planner wants to implement an efficient (cost-minimizing) network satisfying all connection demands.

Recall from section 4.4 that an allocation problem (G, P, C) consists of a connection problem (P, C) (where P is the profile of connection demands for agents in M, and C is the cost matrix of the complete graph with N nodes) and a cost-minimizing connection network G.

Since the planner selects a network based on the reported connection demands and estimated cost matrix, the selected network need not be a MCCN for the true connection problem (in fact, it may not even be a connection network for the true connection problem, if agents misreport their connection demands). We therefore need to define cost-allocation rules ϕ on the domain of allocation problems (G, P, C), where G is any connection network without redundant connections given the problem (P, C) (i.e., a network in the shape of a tree or a forest). Allocation rules ϕ are budget balanced but since there is no equivalent to the irreducible cost matrix for the MCCN problem, we can no longer define relevant rules as reductionist. Consequently, when the planner allocates the true cost of the selected network, the allocation rule will be based on a cost matrix consisting of the true costs for connections in the chosen connection network and on estimated costs for all other connections.

6.3.1 The Setup

Formally, agents report their connectivity demands $\omega = (\omega_1, \ldots, \omega_m)$ with $\omega_i = (\alpha_i, \beta_i) \in E^{N^C}$, as well as all connection costs $\sigma = (\sigma_1, \ldots, \sigma_m)$ with $\sigma_i = (\sigma_i^{jk})_{jk \in E^{NC}}$ in accordance with the cost structure (i.e., $\sigma_i^{jk} > 0$ for all $j \neq k$ and 0 otherwise). As before, the estimation rule τ takes values between min and max reports (i.e., $\tau(\sigma_1^{jk}, \ldots, \sigma_m^{jk}) \in [\min_i\{\sigma_i^{jk}\}, \max_i\{\sigma_i^{jk}\}])$.

We call τ *sensitive* if $\tau(\mu_i^{jk}, \sigma_{-i}^{jk}) > \tau(v_i^{jk}, \sigma_{-i}^{jk})$ for all μ_i^{jk} and v_i^{jk} with $\mu_i^{jk} > v_i^{jk}$. That is, when τ is sensitive, agents can influence the planner's estimate by their reports.

Moreover, we say that τ is *downward* unbounded if $\lim_{\sigma_i^{jk} \to 0} \tau(\sigma_i^{jk}, \sigma_{-i}^{jk}) = 0$. If τ is both downward and upward unbounded (recall the definition in section 6.2.2), then τ is said to be *unbounded*. Indeed, the geometric mean is an example of a sensitive and unbounded estimation rule; the arithmetic mean is sensitive, while the min (max) rule is downward (upward) sensitive as well as downward (upward) unbounded. The median is an example of an estimation rule that is neither sensitive nor unbounded: with such rules, all reports are equally good from every individual agent's viewpoint, since they cannot influence the planner's cost estimate.

Agents' strategies are given by pairs of reported connection demands and costs, $s_i = (\omega_i, \sigma_i)$, and agents' payoffs are given by the negative value of their payment, if the connection demand is satisfied, and $-\infty$ otherwise.

The planner uses the reported connection demands ω as the actual connection demands that must be satisfied by the network, P^ω, and uses the reported costs σ to estimate the cost matrix $C^e(\sigma)$, with $c_{jk}^e(\sigma) = \tau(\sigma_1^{jk}, \ldots, \sigma_m^{jk})$ for all $jk \in E^{N^C}$.

Given the problem $(P^\omega, C^e(\sigma))$, the planner selects a MCCN $G \in \mathcal{M}(P^\omega, C^e(\sigma))$ and allocates the true observed cost based on the matrix $\rho(G, C, C^e)$, with

$$\rho_{jk}(G, C, C^e) = \begin{cases} c_{jk} & \text{for all } jk \in G, \\ c_{jk}^e & \text{otherwise.} \end{cases} \tag{6.3}$$

Since all MCCNs $G \in \mathcal{M}(P^\omega, C^e(\sigma))$ are equally likely to be selected by the planner, each agent i has expected payment

$$E\phi_i(P^\omega, C, C^e(\sigma)) = \frac{\sum_{G \in \mathcal{M}(P^\omega, C^e(\sigma))} \phi_i(G, P^\omega, \rho(G, C, C^e(\sigma)))}{|\mathcal{M}(P^\omega, C^e(\sigma))|} \tag{6.4}$$

for fixed collections of reports (ω, σ). Agents therefore choose strategies $s_i = (\omega_i, \sigma_i)$ to minimize expected payments (6.4).

Example 6.2 (Illustrating the setup)   Consider the graph below with four locations $\{a, b, c, d\}$ and edge costs as illustrated. Ann wants to connect (a, c), and Bob wants to connect (b, d). Thus, the efficient network is $G = \{ad, ab, bc\}$ with a total cost of 4.

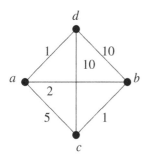

Suppose the planner uses the arithmetic mean as the estimation rule, and the Shapley value (of the game associated with the estimated cost matrix) to allocate costs. Suppose also that both agents report the true connection demands as well as costs. Thus, the planner implements G and allocate the costs as $(-0.5, 4.5)$ to Ann and Bob, respectively. Clearly, this is not a Nash equilibrium: assuming that Ann reports the truth, Bob can gain by misreporting in two ways. First, he can report the true costs, but lie about his connection demand. By reporting that he wants to connect (a, d), the planner will still implement the efficient network G (and therefore Bob's true demand is satisfied anyway), but the total cost is now allocated as $(4, 0)$ instead. Second, Bob can report his true demand, but lie about the costs. By reporting true costs for all edges except bd, where he reports cost 0, the planner will estimate the cost matrix as the true matrix except for $k_{bd}^e = 5$. Thus, the planner will still implement G (so Bob's lie is not revealed), but now costs are split equally $(2, 2)$.

6.3.2 Implementation

Recall the two properties (axioms) unobserved information independence, and network independence (section 4.4.2): the former states that for problems with the same connection structure, but cost structures that differ only outside a common MCCN, agents should pay the same; the latter states that for problems with the same connection and cost structure but different MCCNs, agents should pay the same.

Suppose the planner uses a (continuous) cost-allocation rule violating unobserved information independence (like the Shapley value in example 6.2 above). Then, if everybody reports truthfully, some agent can benefit by making the planner estimate certain costs for connections not included in the chosen MCCN. When the estimation rule τ is continuous and sensitive, this will be possible by reporting some costs for unrealized connections different from their true value (see example 6.2). Moreover, suppose the cost-allocation rule violates network independence. Then, if everybody reports truthfully and there are multiple MCCNs, some agent will prefer that the planner chooses a certain MCCN. When the estimation rule is continuous and sensitive, it will be possible for that agent to make the planner estimate costs such that the desired MCCN becomes the unique MCCN.

In other words, if truthful reporting is a Nash equilibrium, then the cost-allocation rule must satisfy both unobserved information independence and network independence, provided the estimation rule is continuous and sensitive. We record this observation from Hougaard and Tvede (2015) in the following proposition.

Proposition 6.1   Suppose ϕ is continuous in cost structures, and τ is continuous and sensitive. If truthful reporting is a Nash equilibrium, then ϕ satisfies unobserved information independence and network independence.

Recalling theorem 4.3, the planner therefore has no hope of getting the agents to report truthfully with the current four-stage mechanism Φ, unless the announced cost-allocation rule is of the simple form

$$\phi(G, P, C) = \gamma(P, N_C, v(G, C)),$$

where γ maps tuples $(P, N_C, v(G, C))$ to $\mathbf{R}_+^m$. Hougaard and Tvede (2015) prove the following result.[2]

Theorem 6.2   Suppose an allocation rule ϕ satisfies unobserved information independence and network independence, and the estimation rule τ is continuous and sensitive. Then truthful reporting is a Nash equilibrium if and only if γ is increasing in total cost v, and γ_i is independent of ω_i, for all $i \in M$.

Proof   [Sketch of argument]   If γ is increasing in v, and γ_i is independent of ω_i, truthful reporting must be a Nash equilibrium: indeed, given that all other agents report truthfully, no agent can gain by lying about his connection demand (since it has no influence on his payment), and since lying about cost can only increase total cost, no agent stands to gain from this either. Thus, consider the converse claim. Suppose γ_i depends on ω_i, and consider a problem where the unique MCCN satisfies some other agent's connectivity demand by connecting all nodes. Thus i's demand is satisfied independent of her reported demand, and since her payment depends on her report, it may be advantageous to lie. Suppose γ is not increasing in v. Thus, some agent will be favored if the planner implements an inefficient network. Since the estimation rule is continuous and sensitive, agents can misreport such that the desired network is selected by the planner.   ■

Recall the definition of the price of stability (PoS) from chapter 5. By theorem 6.2, worst-case PoS equals 1 when the allocation rule ϕ has the form (4.25) with γ being increasing in v, γ_i independent of ω_i, for every i, and the estimation rule τ is continuous and sensitive.

As noted in section 4.4.3., adding scale invariance to the properties of unobserved information independence and network independence results in the even

---

2. Hougaard and Tvede (2015) also extend the result of theorem 6.2 to a Bayesian setting.

simpler form of cost-allocation rule (4.26) given by

$$\phi(G, P, C) = \delta(P, N_C)v(C, G).$$

Thus, using such a cost-allocation rule in the four-stage mechanism, we have that *truthful reporting is a Nash equilibrium if and only if δ_i is independent of ω_i*, following a similar argument as in theorem 6.2.

Now assume that the planner knows the true connection demands P, but still has to estimate the cost structure, the cost-allocation rule is of the simple form (4.25), and the estimation rule is either downward or upward unbounded. Then it is possible to extend the result from the special case of the MCST model: all Nash equilibria implement an efficient network, that is, $\mathcal{M}(P, C^e(\bar{\sigma})) \subseteq \mathcal{M}(P, C)$ for all $\bar{\sigma} \in \mathcal{NE}$. In other words, if the planner knows connection demands, worst-case PoA equals 1 as well.

However, the following example demonstrates that when the planner is unaware of connection demands, even full knowledge of connection costs does not guarantee implementation of efficient networks, that is, $\mathcal{M}(P^{\bar{\omega}}, C) \cap \mathcal{M}(P, C) = \emptyset$, for some $\bar{\omega} \in \mathcal{NE}$.

Example 6.3 (Inefficient network despite known costs) Consider the following case with four agents $i \in \{A, B, C, D\}$ and four nodes $j \in \{1, 2, 3, 4\}$. All four agents want to connect nodes 2 and 3. The costs are given by $c_{23} = 1$, $c_{12} = c_{34} = \varepsilon > 0$, and $c_{jk} = 1 + \varepsilon$ otherwise, as illustrated by the complete graph below.

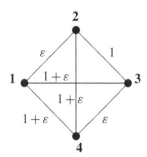

The unique MCCN is $G = \{23\}$ with total cost $v(G, C) = 1$. Suppose that agents A and B report that they want nodes 1 and 3 connected, while agents C and D report that they want nodes 2 and 4 connected. Based on these reports, the cost-minimizing network is given by $G' = \{12, 23, 34\}$ with total cost $v(G', C) = 1 + 2\varepsilon$. So the network is inefficient, and it is not possible for any of the agents to change their payment given the reports of the other agents. Thus, the reported demands constitute a Nash equilibrium, and since ε is arbitrary, the PoA is unbounded (here PoA $= 1 + 2\varepsilon$).

6.3.3 Alternative Game Form

Juarez and Kumar (2013) consider an alternative mechanism inspired by the network cost-sharing games of Anshelevich et al. (2008) and Chen et al. (2010).

1. The planner announces a cost-allocation rule ξ.

2. Every agent i reports, simultaneously and independently, a desired path p_i connecting the demanded locations (a_i, b_i).

3. Based on these reports, the planner implements the network $G = \cup_{i \in M} p_i$ and allocates its cost $C(G)$ using ξ.

It is somewhat difficult to compare this game form directly to the four-stage mechanism Φ analyzed above. In Juarez and Kumar (2013), the scenario resembles the decentralized situation analyzed in chapter 5, where the planner is fully informed but is unable to enforce a centralized solution. So every agent has less impact on the planner's choice of network, since the planner just implements the union of the reported paths. The planner is also better informed, since both connection demands in the form of locations (a_i, b_i) for every agent and all connection costs are known by the planner.

It is assumed that the planner uses a budget-balanced and continuous cost-allocation rule based on minimal information in the sense that ξ is a function only of total network cost $C(\cup_i p_i)$ and individual path costs $C(p_i)$, for all i. That is, $\xi(C(\cup_i p_i); C(p_1), \ldots, C(p_m))$.[3]

With the above game form, an agent's strategy space consists of the set of all paths in the complete graph connecting the desired locations (a_i, b_i); the agent's payoff equals the negative value of the payment, that is, $-\xi_i$. Thus, a profile of paths, p^*, is a Nash equilibrium if every agent's path p_i^* minimizes her payment ξ_i, given the paths chosen by the other agents p_{-i}^*. The rule ξ is called *efficient* if, for any problem, a cost-minimizing network G^* is the Nash equilibrium outcome in the game induced by ξ.

Intuition, based on our analysis so far, indicates that efficient implementation requires that the cost-allocation rule ξ have a particularly simple form. The following example will support this intuition by illustrating the problems

---

3. Since $C(\cup_i p_i) \leq \sum_{i \in M} C(p_i)$, the allocation problem resembles a classic rationing problem (see, e.g., Hougaard, 2009).

of implementing the efficient network using the present game form if payments ξ depend on individual path costs $C(p_i)$.

Example 6.4 (Problems of implementing the efficient network) Consider the graph below. Suppose that Ann wants to connect nodes a and b, while Bob wants to connect nodes c and d. The efficient network consists of the graph $G^* = \{ad, bd, bc\}$ with total (minimum) cost 1.8.

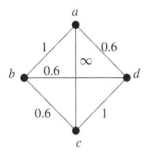

It seems natural to assume that the payment of agent i should be increasing in i's individual path cost $C(p_i)$, but in this case, both agents have incentive to report their direct connection, $p_A = \{ab\}$ and $p_B = \{cd\}$, each with a cost of 1, instead of the more costly paths, $\{ad, bd\}$ and $\{bd, bc\}$, leading to efficient implementation. Thus, the planner has to give agents incentive to prefer paths that may not be individually cost minimizing, and the only relevant way this can be done is by making ξ independent of individual path costs. Thus, if ξ is a monotonic function of total cost (in the sense that no agent benefits from a higher total cost), then reports leading to implementation of an efficient network will be a Nash equilibrium, and any other Nash equilibrium will result in weakly higher payments for all agents. In fact, the reverse observation is also true, as stated below.

A rule ξ is said to *Pareto Nash implement* an efficient network if, for any problem, an efficient network is a Nash equilibrium outcome of the game induced by ξ, and any other network resulting from Nash equilibria implies weakly higher payments for all agents.

Juarez and Kumar (2013) prove that, for $m \geq 3$, the rule ξ Pareto Nash implements the efficient network if and only if there is a monotonic function $f : \mathbb{R}_+ \to \mathbb{R}_+^m$, such that $\sum_{i \in M} f_i(C(G)) = C(G)$ and

$$\xi(C(G); C(p_1), \ldots, C(p_m)) = f(C(G)), \tag{6.5}$$

for any feasible $(C(G); C(p_1), \ldots, C(p_m))$.

So agents' relative cost shares are fixed under a rule that Pareto Nash implements an efficient network. With a rule that uses minimal information, agents are characterized by their individual path cost, so only the equal-split rule satisfies equal treatment of equals in this case. Moreover, note that although PoS equals 1 for allocation rules using fixed relative cost shares, a simple variation of example 6.3 shows that PoA is unbounded.

Example 6.4 (continued: PoA is unbounded)   Recall example 6.3, but consider the case where agents A and B both report the path $p_A = p_B = \{12, 13\}$, while agents C and D both report the path $p_C = p_D = \{24, 34\}$. Therefore, $\overline{G} = \{12, 13, 23, 34\}$ is the implemented network with a total cost of $2 + 4\varepsilon$ versus the total cost of 1 for the efficient network $G = \{23\}$. Reports (p_A, p_B, p_C, p_D) constitute a Nash equilibrium, since deviation by any single agent will only lead to a (weakly) more costly network.

6.4   Welfare-Maximizing Networks

In the connection networks model, it is assumed that all agents' connection demands must be satisfied. Implicitly, the planner therefore construes the problem as if agents have infinite utility from connectivity. Obviously, this is a strong assumption, and it is easy to imagine situations where agents have limited willingness to pay for connectivity. When agents have limited willingness to pay, there may be cases where the cost of connecting certain agents exceeds the welfare gained by society from adding them to the network. From the social planner's viewpoint, an efficient network is therefore not necessarily a cost-minimizing network including all agents, but rather a welfare-maximizing network, which may result in the exclusion of certain agents. It seems natural to define social welfare as the total net benefit of network users (i.e., the network users' total willingness to pay minus the total cost of the network itself).

The agents know the full profile of connection demands and willingness to pay, but the planner does not. We assume that everybody knows connection costs. The present model therefore introduces a new form of tension: agents can ensure being part of the network by indicating a high willingness to pay, but then they are also likely to pay a high price for connectivity. Conversely, by indicating a low willingness to pay, their payments may be low, but they risk being excluded from the network.

In terms of implementation, extending the connection networks model to include agents' willingness to pay is not without problems. The example below shows that using the four-stage game form Φ, with the obvious adaptation to the extended setting, may lead to situations without Nash equilibria.

Example 6.5 (No Nash equilibria) Assume the planner knows connection costs and connection demands, but is unaware of every agent's willingness pay w_i. In contrast, agents are fully informed. A straightforward adaptation of the kind of game form, Φ, analyzed above is as follows. (1) The planner announces an estimation rule and a cost sharing rule. (2) Every agent reports, simultaneously and independently, the profile of willingness to pay for all agents. (3) Based on these reports, the planner estimates the willingness to pay for every agent and selects at random a welfare-maximizing network. (4) The planner allocates the cost of the selected network according to the announced cost-sharing rule.

Now, consider four locations $\{a, b, c, d\}$ and costs $c_{ab} = 4$, $c_{ac} = c_{bd} = 1$, $c_{cd} = 3$, and ∞ otherwise, as illustrated by the graph below.

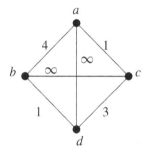

Suppose Ann has connection demand (a, b), and Bob has connection demand (c, d). Moreover, their (true) willingness to pay is given by $w_A = 3.1$ and $w_B = 2.1$, respectively. While it generates negative welfare to satisfy the connection demand of either Ann or Bob individually, there is a total welfare gain of $3.1 + 2.1 - 5 = 0.2$ from connecting both of them in the welfare-maximizing network $G^* = \{ac, cd, bd\}$ (with a total cost $v(G^*) = 5$).

Suppose the planner sets up the game form Φ, as described above, using the average as estimation rule and the constraint equal split as a cost-sharing rule; that is, $\phi_i = \min\{w_i, \rho\}$ with $\phi_A + \phi_B = v(G)$. (Note that equal split is not relevant, since no agent should pay more than their willingness to pay, as in voluntary participation). The question is whether this mechanism can implement the welfare-maximizing network G^*. Suppose Ann reports the willingness to pay profile $(2.9, 2.1)$. Then the best response of Bob would be to report $(3.3, 1.7)$, yielding optimal network G^* and payments $\phi = (3.1, 1.9)$ (equal to the planners estimation of their willingness to pay). However, the best response of Ann would then be to report $(2.5, 2.5)$, yielding payments $\phi = (2.9, 2.1)$, and so forth. Thus, there is no Nash equilibrium, since the two agents can never agree on how share the total welfare gain of 0.2.

At first glance, one could therefore doubt that there are ways to implement welfare-maximizing networks. However, even full implementation for Nash equilibrium can be obtained at the price of distributional fairness.

6.4.1   The Setup

Following Hougaard and Tvede (2017), we let $M = \{1, \ldots, m\}$ be a set of finitely many agents, with $m \geq 3$, and $\mathcal{L}$ be a set of finitely many locations. The set of connections between pairs of locations is $\mathcal{L}^2 = \mathcal{L} \times \mathcal{L}$. As in the MCCN model, the *cost structure* C describes the costs of all edges in the complete graph $G(N_C, E^{N_C})$ with $N_C \subseteq \mathcal{L}$, where $c_{jj} = 0$ for all $j \in N_C$, and $c_{jk} > 0$ for all $j, k \in N_C$ where $j \neq k$. Since the network is undirected, $c_{jk} = c_{kj}$. Connections (edges) are assumed to be public goods.

Every agent $i \in M$ has a connection demand in the form of a pair of locations $(a_i, b_i) \in \mathcal{L}^2$ that she wants to be directly or indirectly connected, and a willingness to pay $w_i > 0$ if her connection demand is satisfied. A *demand structure* D is a collection of individual connection demands and willingness to pay, $(D_i)_{i \in M}$, with $D_i = (a_i, b_i, w_i)$ for every i. Let $\mathcal{D}$ denote the set of demand structures.

Let $\mathcal{G}$ denote the set of graphs on $\mathcal{L}$. Recall that for a given graph $G \in \mathcal{G}$, a *path* between locations (nodes) a and b is a finite sequence of connections (edges) $p_{ab} = ah_1, h_1h_2, \ldots, h_kb$ in G. Let P_i be the set of paths between a_i and b_i in the complete graph G^N. For a graph and a demand structure (G, D), let $\mathcal{I}(G, D)$ be the set of agents i whose connection demands are satisfied in G (i.e., $i \in \mathcal{I}(G, D)$ if and only if $P_i \cap G \neq \emptyset$). For a graph G, let $v(G) = \sum_{jk \in G} c_{jk}$ be the total cost of G.

For a graph and a demand structure (G, D), the *social welfare* is given by

$$\sum_{i \in \mathcal{I}(G,D)} w_i - v(G).$$

A graph G is a *welfare-maximizing network*, for demand structure D, if and only if

$$\sum_{i \in \mathcal{I}(G,D)} w_i - v(G) \geq \sum_{i \in \mathcal{I}(G',D)} w_i - v(G')$$

for every graph $G' \in \mathcal{G}$. The set of welfare-maximizing networks is nonempty and finite, because the set of graphs is nonempty and finite. Let $\mathcal{WMN}(D)$ be the set of welfare-maximizing networks for demand structure D. Clearly every such network is either a tree or a forest (a collection of trees). Indeed if a graph contains a cycle, then removing any connection in the cycle does not change whether connection demands are satisfied or not. Since every welfare-maximizing network is either a tree or a forest, for every agent i, there is a unique path $p_i(G, D)$ between a_i and b_i in G.

An *outcome* is a graph and an allocation of its total cost, (G, ϕ^G), where $\sum_i \phi_i^G = v(G)$. Let $\mathcal{O}$ be the set of outcomes.

Let $\mathcal{O}^* \subset \mathcal{O}$ be the set of *desirable outcomes*, that is, outcomes $(G, \phi^G) \in \mathcal{O}$ for which $G \in \mathcal{WMN}(D)$, and $\phi_i^G \le w_i$, for every $i \in \mathcal{I}(G, D)$, and $\phi_i^G = 0$ for every $i \notin \mathcal{I}(G, D)$. That is, an outcome is desirable if G is a welfare-maximizing network, no agent included in the network pays more than their willingness to pay, and agents excluded from the network pay nothing.

A *solution* $\Gamma : \mathcal{D} \to \mathcal{O}$ is a mapping from demand structures to outcomes. In particular, we are interested in the solution correspondence Γ^M for which $(G, \phi^G) \in \Gamma^M$ if and only if $(G, \phi^G) \in \mathcal{O}^*$. Even though desirable outcomes ensure welfare-maximizing networks satisfying individual rationality, the distribution of the obtained welfare can be very unequal, for instance, allocating the entire welfare gain to a single agent.

Given a demand structure $D \in \mathcal{D}$ and an outcome $(G, \phi^G) \in \mathcal{O}$, let the payoff of every agent $i \in M$ be defined by

$$u_i(G, \phi^G, D) = \begin{cases} w_i - \phi_i^G & \text{for } P_i \cap G \neq \emptyset, \\ -\phi_i^G & \text{for } P_i \cap G = \emptyset. \end{cases}$$

That is, if agent i is included in the network, the payoff equals his net gain (i.e., willingness to pay minus payment), and if i is excluded from the network then the payoff equals minus the payment.

In the following, let us focus on the potential for implementation in Nash equilibrium: I do not suggest specific game forms, but such game forms may be constructed on the basis of Maskin's canonical mechanism (Maskin, 1999). To analyze the potential for Nash implementation, we need to review a few concepts from classic implementation theory in the present context.

6.4.2 Maskin Monotonicity

A solution Γ is said to be *Nash implementable* provided there exists a game form $\overline{\Phi} = ((S_i)_i, f)$, where S_i is the strategy set of agent i, and f is a map from strategy profiles to outcomes, $f : \times_i S_i \to \mathcal{O}$, such that for all demand structures D, the set of Nash equilibria for $\overline{\Phi}$ is $\Gamma(D)$. In other words, Γ is Nash implementable if, for any demand structure D, every outcome $(G, \phi^G) \in \Gamma(D)$ is a Nash equilibrium for $\overline{\Phi}$, and every Nash equilibrium outcome of $\overline{\Phi}$ is an outcome in $\Gamma(D)$.

For a given agent $i \in M$, demand structure $D \in \mathcal{D}$, and outcome $(G, \phi^G) \in \mathcal{O}$, let

$$L_i(G, \phi^G, D) = \{(G', \phi^{G'}) \in \mathcal{O} \mid u_i(G', \phi^{G'}, D) \leq u_i(G, \phi^G, D)\}$$

be the set of outcomes for which agent i prefers (G, ϕ^G) given demand structure D (i.e., agent i's *lower contour set* at (G, ϕ^G, D)).

A solution Γ is *Maskin monotonic* provided that for all $(G, \phi^G) \in \mathcal{O}$ and $D, D' \in \mathcal{D}$ such that $(G, \phi^G) \in \Gamma(D)$, then $L_i(G, \phi^G, D) \subset L_i(G, \phi^G, D')$ for every i implies $(G, \phi^G) \in \Gamma(D')$. In other words, Maskin monotonicity requires that if an outcome (G, ϕ^G) is a solution for demand structure D, and if (G, ϕ^G) does not fall in the ranking of any agent when the demand structure changes to D', then (G, ϕ^G) is also a solution for demand structure D'. In this sense, Maskin monotonicity is a solution-invariance condition with respect to monotone changes in demand structure.

In our context, Maskin (1999) shows that *a solution is Nash implementable if and only if it is Maskin monotonic*. The argument showing that Maskin monotonicity implies Nash implementability is constructive, so we present a sketch related to the present context.

We consider the following canonical game form: a strategy for each agent consists of an outcome, a demand structure, and an integer, i.e., each agent, $i \in M$, reports $s_i = ((G, \phi^G)^i, D^i, n^i)$. Define the mechanism for the outcome function f as follows.

1. If $s_i = ((G, \phi^G)^i, D^i, n^i) = ((G, \phi^G), D, n) = s$ for all $i \in M$ and $(G, \phi^G) \subset \Gamma(D)$, then $f(s) = (G, \phi^G)$. In words, if agents submit identical reports, and the reported outcome is a solution, then the mechanism chooses the reported outcome.

2. Suppose there exists $j \in M$ such that $((G, \phi^G)^i, D^i, n^i) = ((G, \phi^G), D, n)$ for all $i \neq j$, and $(G, \phi^G) \in \Gamma(D)$, but $s_j \neq ((G, \phi^G), D, n)$. Then $f(s) = (G, \phi^G)^j$ if $(G, \phi^G)^j \in L_j(G, \phi^G, D)$, and $f(s) = (G, \phi^G)$ otherwise. In words, if all but one agent, j, agree in their reports, the mechanism chooses the outcome reported by j only if this is worse for j than the outcome reported by the $m - 1$ other agents (given their reported demand structure). Otherwise, the mechanism chooses the majority outcome.

3. In all other cases, the mechanism chooses the outcome $f(s) = (G, \phi^G)^i$ reported by the agent with the highest reported integer n^i.

Now, for any $(G, \phi^G) \in \Gamma(D)$, it is a Nash equilibrium for all agents to report $((G, \phi^G), D, 1)$. Indeed, by construction of f, item 2 in the above mechanism makes sure that no individual can gain by deviation.

We now need to check that all Nash equilibria result in outcomes $(G, \phi^G) \in \Gamma(D)$. A Nash equilibrium can not appear under item 3, since whatever outcome the mechanism has selected, some agent will prefer an outcome where he pays less and is able to enforce that by reporting a higher integer (note that cost shares ϕ_i^G may be negative).

A Nash equilibrium also cannot appear under item 2, since any agent $i \neq j$ can enforce his preferred outcome by reporting an integer higher than the others (because item 3 applies in that case).

So we are left with the case where $s_i = s = ((G, \phi^G), D', n)$ for all $i \in M$, with $D' \neq D$ and $(G, \phi^G) \in \Gamma(D')$. Any agent can deviate and obtain an outcome (H, ϕ^H), where $(H, \phi^H) \in L_i(G, \phi^G, D')$ (because item 2 applies in that case). Thus, for s to be a Nash equilibrium, we must have that $(H, \phi^H) \in L_i(G, \phi^G, D')$ implies $(H, \phi^H) \in L_i(G, \phi^G, D)$, for every $i \in M$. So by Maskin monotonicity, $(G, \phi^G) \in \Gamma(D)$.

6.4.3 Implementation Results

Hougaard and Tvede (2017) prove that the correspondence of desirable outcomes is Nash implementable.

Theorem 6.3 The solution Γ^M is Nash implementable.

Proof [Sketch of argument] By Maskin (1999), we need to show that Γ^M is Maskin monotonic. Let $\sigma = (G, \phi^G) \in \Gamma^M(D)$, and $\tau = (H, \phi^H) \in \mathcal{O}$. For every agent $i \in M$, the conditions under which outcome τ is worse than outcome σ (i.e., $\tau \in L_i(\sigma, D)$) depends on whether connection demands are satisfied:

1. Suppose $G, H \cap P_i = \emptyset$, then $\tau \in L_i(\sigma, D)$ if and only if $\phi_i^H \geq 0$.

2. Suppose $G \cap P_i = \emptyset$ and $H \cap P_i \neq \emptyset$, then $\tau \in L_i(\sigma, D)$ if and only if $w_i - \phi_i^H \leq 0 \Leftrightarrow \phi_i^H \geq w_i$.

3. Suppose $G \cap P_i \neq \emptyset$ and $H \cap P_i = \emptyset$, then $\tau \in L_i(\sigma, D)$ if and only if $0 \leq w_i - \phi^G \Leftrightarrow w_i \geq \phi_i^G$.

4. Suppose $G, H \cap P_i \neq \emptyset$, then $\tau \in L_i(\sigma, D)$ if and only if $w_i - \phi_i^H \leq w_i - \phi^G \Leftrightarrow \phi_i^G \leq \phi_i^H$.

First, consider two demand structures $D, D' \in \mathcal{D}$, for which only the willingness to pay differs, but the connection demands do not, that is, $P_i = P_i'$.

In the case where agent i is excluded from the network G (i.e., $G \cap P_i = \emptyset$), we have $L_i(\sigma, D) \subset L_i(\sigma, D')$ if and only if $w_i' \leq w_i$ (indeed, when decreasing the willingness to pay, more outcomes become worse than σ).

In the case where agent i is part of the network G (i.e., $G \cap P_i \neq \emptyset$), we have $L_i(\sigma, D) \subset L_i(\sigma, D')$ if and only if $w_i' \geq w_i$ (indeed, when increasing the willingness to pay, more outcomes become worse than σ).

Consequently, when $L_i(\sigma, D) \subset L_i(\sigma, D')$ for every i, all agents excluded by σ with D will have lower willingness to pay with D', and those who are part of the network with D will have higher willingness to pay with D': that is, $\sigma \in \Gamma^M(D')$.

Next, consider two demands structures $D, D' \in \mathcal{D}$, where connection demands may also differ, that is $P_i \neq P_i'$.

In the case where agent i is excluded with D, but is part of the network with D', we clearly have $L_i(\sigma, D) \subset L_i(\sigma, D')$. In the three other cases, however, we can show that there will exist some i for which $L_i(\sigma, D) \not\subset L_i(\sigma, D')$. Consequently, monotonicity is trivially satisfied in these cases. ∎

However, even if the planner knows the profile of connection demands $(a_i, b_i)_i$ (but not the willingness to pay for every agent), no selection from the set of desirable outcomes can be Nash implemented, as shown by the following example.

Example 6.5 (continued: No selection from Γ^M can be Nash implemented)
Recall the case from example 6.5 above, with four locations $\{a, b, c, d\}$ and costs $c_{ab} = 4$, $c_{ac} = c_{bd} = 1$, $c_{cd} = 3$, and ∞ otherwise. Assume $m - 1$ agents (considered as one aggregate agent i) have connection demands (a, b), and one agent, j, has connection demand (c, d). First, consider the demand structure D, where $w_i = 3.9$ and $w_j = 1.2$. In this case the welfare-maximizing network becomes $G = \{ac, cd, bd\}$ with a total cost $v(G) = 5$. Thus, $\sigma = (G, \phi_i, \phi_j)$ with $\phi_i \in [3.8, 3.9]$, $\phi_j \in [1.1, 1.2]$, and $\phi_i + \phi_j = 5$ is a potential selection from $\Gamma^M(D)$. Next, consider the demand structure $\tilde{D}$, where $\tilde{w}_i = 3.7$ and $\tilde{w}_j = 1.4$. G is still a welfare-maximizing network, so $\tilde{\sigma} = (G, \tilde{\phi}_i, \tilde{\phi}_j)$ with $\tilde{\phi}_i \in [3.6, 3.7]$, $\tilde{\phi}_j \in [1.3, 1.4]$, and $\tilde{\phi}_i + \tilde{\phi}_j = 5$ is a potential selection from $\Gamma(\tilde{D})$. Finally, consider the demand structure D', where $w_i' = \max\{w_i, \tilde{w}_i\} = 3.9$ and $w_j' = \max\{w_j, \tilde{w}_j\} = 1.4$. Since $w_i' = w_i > \tilde{w}_i$ and $w_j' = \tilde{w}_j > w_j$, we have that $L(\sigma, D) \subseteq L(\sigma, D')$ for every agent, and that $L(\tilde{\sigma}, \tilde{D}) \subseteq L(\tilde{\sigma}, D')$ for every agent. Thus, if the selection γ is Maskin monotonic, then we have $\sigma, \tilde{\sigma} \in \gamma(D')$, contradicting that γ is a selection from Γ^M, since $\sigma \neq \tilde{\sigma}$. Thus, no selection γ is Maskin monotonic, and consequently, no selection can be Nash implemented.

So the planner cannot consistently use a specific cost-allocation rule for every problem but has to be flexible when allocating costs. We also have to accept that the distribution of welfare can be very unequal. In fact, it is easy to show that no Nash implementable solution can guarantee positive welfare gains to all agents (when the total welfare is positive).

Example 6.6 (No guarantee of positive payoff) Consider a case where a welfare-maximizing network G connects all agents, but every individual payoff is zero, that is, $w_i = \phi_i^G$ for all $i \in M$. Clearly, $(G, \phi^G) \in \Gamma^M(D)$. Now consider a new demand structure D', where $(a_i', b_i') = (a_i, b_i)$ and $w_i' \geq w_i$ for all $i \in M$ (with at least one strict inequality). Thus, $L_i(G, \phi^G, D) \subseteq L_i(G, \phi^G, D')$ for all $i \in M$. So by Maskin monotonicity, $(G, \phi^G) \in \Gamma^M(D')$. Yet if only a single agent, say j, has higher willingness to pay in D' than in D, the total welfare gain $\Delta = w_j' - w_j$ can only be allocated to agent j if (G, ϕ^G) must be a solution for D'. In other words, allocating a small share of Δ to any other agent $i \neq j$ violates monotonicity.

In Hougaard and Tvede (2017) it is further shown that Γ^M is the unique solution that is implementable for strong Nash equilibrium.

6.5 Subscription Mechanisms

The necessary and sufficient conditions for implementation in SPNE are weaker than those of Nash implementation (see, e.g., Moore and Repullo, 1988; Abreu and Sen, 1990). Therefore, Mutuswami and Winter (2002) are able to show that a selection from the correspondence Γ^M can be implemented in SPNE. That is, a welfare-maximizing network can be obtained for every SPNE of a game form based on sequential reporting by the agents. Furthermore, the resulting payoff is uniquely determined and ensures individual rationality.

6.5.1 The Setup

Mutuswani and Winter (2002) consider a network formation setting à la Jackson and Wolinsky (1996), but with asymmetric information: agents are fully informed about private benefit functions and connection costs, while the planner only knows about connection costs.

Let $N = \{1, \ldots, n\}$ be a finite set of agents. Agents are here identical to nodes, and edges ij in a graph $G \subseteq G^N$ indicate a bilateral relationship between agents i and j. Let G_S denote the set of graphs involving connections between members of $S \subseteq N$ only. Given a graph G, an agent i is said to be *excluded* if there exists no agent j such that $ij \in G$. The cost of establishing a network (graph $G \subseteq G^N$) is given by a nonnegative cost function $C : G^N \to \mathbb{R}$, for which $C(\emptyset) = 0$.

For every agent $i \in N$, the net payoff obtained by i in network G is given by the quasilinear function $u_i(G, x_i) = v_i(G) - x_i$, where v_i is a nonnegative individual benefit function that is assumed to be monotonic (in the sense that $G \subseteq G' \Rightarrow v_i(G) \leq v_i(G')$), and x_i is the payment (cost share) of agent i.

Denote by S_k, for $k = 1, \ldots, n$, the upper coalition of agents $\{k, \ldots, n\}$ (given the order $1, \ldots, n$).

6.5.2 Two Game Forms

In the first game form, Φ_n, agents report sequentially in a given order, say, $1, \ldots, n$:

- When it is agent i's turn, i reports a tuple $\sigma_i = (G_i, x_i)$, where G_i is the graph (set of connections) that i wants to see formed, and x_i is her willingness to pay if the implemented network includes G_i, (i.e., $G \supseteq G_i$).

- Given the profile of reports σ, the planner selects the largest coalition among $\{\emptyset, S_1, \ldots, S_n\}$ such that $G_i \in G_S$, for all $i \in S$, and $\sum_{i \in S} x_i \geq C(\cup_{i \in S} G_i)$. Denote by S^* the largest such coalition, and let $G^* = \cup_{i \in S^*} G_i$ be the network that is implemented.

- Agents pay x_i if $i \in S^*$, and 0 otherwise.

The reason that agents report sequentially, while the planner considers the entire profile of reports becomes clear when analyzing the mechanism below. Notice that G^* may include several components, and that agents in S^* propose graphs that only connect agents in S^*. In other words, G^* excludes the complement of S^*. Moreover, notice that x_i corresponds to the *announced* willingness to pay, while the true willingness to pay can be interpreted as $v_i(G_i)$. Given the order of announcements, agents may benefit from lying about their willingness to pay.

In practice, sequential reporting seems somewhat unfortunate: it creates an order dependance with the unfairness that follows (see e.g., example 6.5 above); and it takes time to execute, if N is large and each agent needs a given time interval to respond. The former problem can be partly solved by a second version of the mechanism defined below.

The second game form, Φ_n', builds on the first:

- Agents play Φ_n according to an arbitrarily chosen order (i.e., with equal probability of any order).

- Agents are asked whether they want to replay Φ_n.

- The game ends when no agent wants a replay.

Intuitively, the induced game will stop after the first step if agents obtain their expected payoffs over all orderings and discount future payoffs.

6.5.3   Implementation Results

Define the *standalone* payoff for coalition $S \subseteq N$, $S \neq \emptyset$ as

$$sa(S) = \max_{G \in G_S} \left(\sum_{i \in S} v_i(G) - C(G) \right), \tag{6.6}$$

and set $sa(\emptyset) = 0$. Clearly, a graph G solving $sa(N)$ is a welfare-maximizing network for N. There may be multiple such graphs for a given problem. Now, for every agent $i \in N$, define the *marginal contribution* of agent i, for the order $1, \ldots, n$, as

$$u_i^* = sa(S_i) - sa(S_{i+1}), \tag{6.7}$$

with $S_{n+1} = \emptyset$. Since $S_i \supseteq S_{i+1}$ we have that $u_i^* \geq 0$ for all i: Indeed, this follows because a welfare-maximizing network for a coalition S can always be formed by a larger coalition $T \supseteq S$. Thus $sa(T) \geq sa(S)$.

Moreover, let $\phi^{Sh}(sa) = (\phi_1^{Sh}(sa), \ldots, \phi_n^{Sh}(sa))$ denote the profile of Shapley payoffs in the game (N, sa). Recall from exercise 2.8, that the core of (N, sa) may be empty.

Since both mechanisms induce a dynamic game, the natural equilibrium concept is SPNE. Mutuswami and Winter (2002) now prove the following result.

Theorem 6.4   All SPNEs of the mechanism Φ_n result in a welfare maximizing network, and agents' net payoffs in all SPNEs are given by $(u_1^*, \ldots, u_n^*)$. Moreover, all SPNEs of the mechanism Φ_n' result in a welfare-maximizing network, and agents' net payoffs in all SPNEs are given by $\phi^{Sh}(sa)$.

Proof [Sketch of argument]   First consider the mechanism Φ_n. We need to show that if u is a profile of net payoffs in some SPNE, we must have $u \geq u^*$. Indeed, say $u_k < u_k^*$ for some agent k, then k can deviate by reporting (G_k', x_k'), where G_k' is welfare-maximizing for coalition S_k, and x_k' is determined such that $sa(S_{k+1}) - (\sum_{j=k+1}^n v_j(G_k') - c(G_k')) < x_k' < v_k(G_k') - u_k$. It can be shown that such a deviation is profitable for agent k, hence contradicting that we had an equilibrium in the first place. Since $\sum_{i \in N} u_i^* = sa(N)$, we must have $u = u^*$ in equilibrium.

Next, consider the mechanism Φ_n'. Since the resulting net payoffs in any SPNE are given by u^* in the first step, and since each order is equally likely, the expected net payoff at the start of second step is ϕ^{Sh}. Thus, if any agent i

ends up getting less than ϕ_i^{Sh} after first step, she will force the game to continue by accepting replay in the second step. Thus, if G^* is the welfare-maximizing network for N, the strategy profile $(\{G^*, x_i^*\}, \text{"}No\text{"})_{i=1}^n$, with $x_i^* = v_i(G^*) - \phi_i^{Sh}$, is SPNE. For G^*, this SPNE is furthermore unique if agents discount future payoff or have lexicographic preferences for the game ending after the first step. ∎

So there may be multiple SPNEs (for instance, when multiple welfare-maximizing networks exist for a given problem), but the equilibrium net payoff is uniquely determined by u^* for a given order, and ϕ^{Sh} in expectation. With a welfare-maximizing network G^*, we have that $v_i(G^*) - x_i = u_i^* \geq 0$ for every agent i. Thus, no agent included in G^* pays more than his resulting benefit. Moreover, note that the price of improving equity when going from Φ_n to Φ_n' comes in the form of computational complexity of equilibrium strategies, determining the Shapley value when N is large.

Example 6.6 (continued: Using mechanisms Φ_n and Φ_n') Recall example 6.5 above, where no Nash equilibria existed under the game form Φ. We will now demonstrate that instead using either game form Φ_n or Φ_n' results in efficient implementation. We here interpret Ann's benefit as $v_A(G) = v_A(G^*) = 3.1$ for all $G \supseteq G^*$, and $v_A = 0$ otherwise. Similarly, Bob's benefit is $v_B(G) = v_B(G^*) = 2.1$ for all $G \supseteq G^*$, and $v_B = 0$ otherwise. Thus, G^* remains the welfare-maximizing network, and the induced game (N, sa) is given by $sa(Ann) = sa(Bob) = 0$; and $sa(Ann, Bob) = 0.2$.

Consider the mechanism Φ_n, and the order of reporting: Ann first, Bob second. The marginal contributions become $u^* = (0.2, 0)$ for Ann and Bob, respectively, so Ann reporting $(G^*, 2.9)$ and Bob reporting $(G^*, 2.1)$ constitute an SPNE. Indeed, by backward induction, Bob's best reply is $(G^*, 5 - x_A)$ if $v_B(G^*) - 5 + x_A \geq 0$ and $(\emptyset, 0)$ otherwise. Thus, it is optimal for Ann to report $(G^*, 2.9)$. If we reverse the order of reporting, so Bob reports first and Ann second, the marginal contributions become $u^* = (0, 0.2)$ for Ann and Bob, respectively. Thus, Bob reporting $(G^*, 1.9)$ and Ann reporting $(G^*, 3.1)$ constitutes an SPNE in this case.

Using the mechanism Φ_n' we get that the reports $(G^*, 3)$ and $(G^*, 2)$ from Ann and Bob, respectively, constitute an SPNE with expected net payoff $\phi^{Sh} = (0.1, 0.1)$.

Since agents' benefits from being included in the network are nonnegative, it seems unfortunate that some equilibria can be supported by a reported willingness to pay that is negative, and hence not trustworthy. This can happen even in case of the game form Φ'_n. By a variation of example 1 in Mutuswami and Winter (2002), we get the example below.

Example 6.7 (Agents may report negative willingness to pay in SPNE of Φ'_n)
Consider three agents $N = \{1, 2, 3\}$. For $j = 1, 2, 3$, let $G^j = \{ij, jk\}$, where $k \neq i$, $j \neq k$, and $i \neq j$. Now, agents' benefit functions are given by

$v_1(\{1, j\}) = 10$ for $j = 2, 3$; $v_1(G^j) = 36$ for all j; $v_1(G^N) = 40$;

$v_2(\{2, j\}) = 16$ for $j = 1, 3$; $v_2(G^j) = 26$, for all j; $v_2(G^N) = 30$;

$v_3(\{3, j\}) = 16$ for $j = 1, 2$; $v_3(G^j) = 16$ for all j; $v_3(G^N) = 20$.

Excluded agents get 0. Note that v is monotonic for all agents.

Suppose that the cost of establishing any connection ij is 15. Then the stand-alone values are given by $sa(\{i\}) = 0$ for all i, $sa(\{1, 2\}) = sa(\{1, 3\}) = 11$, $sa(\{2, 3\}) = 17$, and $sa(N) = 48$. Any of the graphs G^j constitutes an efficient network. Using the game form Φ'_n the net payoffs in any SPNE are given by the Shapley value of the induced game (N, sa), which in the above case is $(14, 17, 17)$. Thus, in equilibrium the reported willingness to pay of the three agents is given by $x = (22, 9, -1)$. It does not seem reasonable that agent 3 should be successful in reporting a negative willingness to pay.

6.6 Exercises

6.1. Recall the case in example 6.1. Is truthful reporting a Nash equilibrium using the game form Φ with estimation rule $\tau(\sigma_1^{jk}, \ldots, \sigma_n^{jk}) = \max\{\sigma_1^{jk}, \ldots, \sigma_n^{jk}\}$, for all jk, and the cost allocation rule $\phi_i = \frac{k_{0i}^e}{\sum_j k_{0j}^e} v(G, C)$? What if the planner uses the minimum as the estimation rule instead?

Finally, comment on the following mechanism, and compare it to Φ. The planner selects an arbitrary agent, who is asked to report all connection costs. Independent of what is reported, the selected agent pays nothing when the total network cost is allocated.

6.2. (Hougaard and Tvede, 2015) Present a formal proof of proposition 6.1.

6.3. (Juarez and Kumar, 2013) Prove that if a cost-allocation rule ξ using minimal information is efficient, then it is monotonic in total cost. That is, if total cost increases (ceteris paribus), then no agent gets a lower cost share.

6.4. Consider the game form of Juarez and Kumar (2013). Find an example showing that $\xi_i = \frac{C(p_i)}{\sum_j C(p_j)} C(\cup_j p_j)$ does not Pareto Nash implement the efficient network. Now, define agent i's standalone cost as the minimal cost of satisfying i's demand (i.e., $SA_i = \min_{p_i} C(p_i)$), and let $\bar{\xi}_i = \frac{SA_i}{\sum_j SA_j} C(\cup_j p_j)$. Does $\bar{\xi}$ use minimal information? Does $\bar{\xi}$ Pareto Nash implement the efficient network? Try to define what it means to implement the efficient network in strong Nash equilibrium. Does $\bar{\xi}$ implement the efficient network in strong Nash equilibrium?

6.5. Consider implementation in the MCST-model. Given a cost matrix K, an outcome $(G, \phi^G(K))$ consists of a spanning tree G and an allocation of its cost, with $\sum_{i \in N} \phi_i^G(K) = v(G, K)$. Denote by $\mathcal{O}^e$ the set of efficient outcomes (i.e., outcomes where G is a MCST, given K). Now define a solution Γ by $(G, \phi^G(K)) \in \Gamma(K)$ if and only if $(G, \phi^G(K)) \in \mathcal{O}^e$.

Show that if ϕ^G is reductionist and monotonic, then Γ is Maskin monotonic and consequently Nash implementable.

6.6. Recall the case in example 6.5. Will the result of the analysis change if agents only report their own willingness to pay? If yes, what are the Nash equilibria?

Also, recall the mechanism suggested in exercise 6.1, where the planner selects an arbitrary agent, who is asked to report the full profile of connection demands and willingness to pay; independent of what is reported, the selected agent pays nothing when the total network cost is allocated. Will such a mechanism implement a welfare-maximizing network?

6.7. (Hougaard and Tvede, 2017) Consider the simple case with three nodes $\{a, b, c\}$: Ann wants to connect (a, b), and Bob wants to connect (a, c). Edge costs are given by $c_{ab} = 3$, and $c_{ac} = c_{bc} = 2$. Show how a welfare-maximizing network depends on the size of the willingness to pay for Ann (w_A) and Bob (w_B). Try to construct a smallest Nash implementable solution in this case (i.e., the smallest solution satisfying Maskin monotonicity).

6.8. Consider the solution Γ^M. Suppose that, for the true demand structure D, a single agent, j, has a willingness to pay w_j larger than the cost of her connecting path $p_j(G, D)$. Let $\tilde{D}$ be given by $(\tilde{a}_i, \tilde{b}_i) = (a_i, b_i)$, and $\tilde{w}_i = 0$ for all $i \in M$. Note that $(\emptyset, \{0\}) \in \Gamma^M(\tilde{D})$. Explain why $s_i = ((\emptyset, \{0\}), \tilde{D}, 1)$, for all $i \in M$, cannot be a Nash equilibrium in Maskin's canonical mechanism.

Moreover, consider a solution $\tilde{\Gamma}$, where $(G, \pi^G) \in \tilde{\Gamma}$ if and only if $G \in \mathcal{WMN}(D)$ and $\pi_i^G \leq w_i$ for all $i \in M$ (note that $\tilde{\Gamma} \supseteq \Gamma^M$; explain why). Suggest a game form that implements $\tilde{\Gamma}$ in strong Nash equilibrium.

6.9. Adapt your answer to exercise 2.8 to the setup of Mutuswami and Winter (2002) in section 6.4.2. Find SPNEs using both mechanisms Φ_n and Φ'_n, and illustrate that they are vulnerable to coalitional deviations.

6.10. (Mutuswami and Winter, 2002). Following up on exercise 6.9, we say that a function $f : G^N \to \mathbb{R}$ is *supermodular [submodular]* if

$$f(G) + f(G') \leq [\geq] f(G \cup G') + f(G \cap G').$$

Try to prove that if the benefit function v_i is supermodular and the cost function C is submodular, then the game (N, sa) is convex. Combine this result with theorem 6.4 to find out when all SPNEs from Φ_n and Φ'_n are coalitionally stable.

References

Abreu, D., and A. Sen (1990), "Subgame Perfect Implementation: A Necessary and Sufficient Condition," *Journal of Economic Theory* 50: 285–299.

Anshelevich, E., A. Dasgupta, J. Kleinberg, E. Tardos, T. Wexler, and T. Roughgarden (2008), "The Price of Stability for Network Design with Fair Cost Allocation," *SIAM Journal of Computing* 38: 1602–1623.

Bogomolnaia, A., R. Holzman, and H. Moulin (2010), "Sharing the Cost of a Capacity Network," *Mathematics of Operations Research* 35: 173–192.

Chen, H.-L., T. Roughgarden, and G. Valiant (2010), "Designing Network Protocols for Good Equilibria," *SIAM Journal of Computing* 39: 1799–1832.

Hougaard, J. L. (2009), *An Introduction to Allocation Rules.* New York: Springer.

Hougaard, J. L., and M. Tvede (2012), "Truth-Telling and Nash Equilibria in Minimum Cost Spanning Tree Models," *European Journal of Operational Research* 222: 566–570.

Hougaard, J. L. and M. Tvede (2015), "Minimum Cost Connection Networks: Truth-Telling and Implementation," *Journal of Economic Theory* 157: 76–99.

Hougaard, J. L., and M. Tvede (2017), "Implementation of Welfare-Maximizing Connection Networks," manuscript.

Jackson, M. O. (2001), "A Crash Course in Implementation Theory," *Social Choice and Welfare* 18: 655–708.

Jackson, M. O., and A. Wolinsky (1996), "A Strategic Model for Social and Economic Networks," *Journal of Economic Theory* 71: 44–74.

Juarez, R., and R. Kumar (2013), "Implementing Efficient Graphs in Connection Networks," *Economic Theory* 54: 359–403.

Maskin, E. (1999), "Nash Equilibrium and Welfare Optimality," *Review of Economic Studies* 66: 23–38.

Maskin, E., and T. Sjöström (2002), "Implementation theory," in K. J. Arrow et al. (eds.), *Handbook of Social Choice and Welfare*, vol. 1. Amsterdam: North-Holland.

Moore, J., and R. Repullo (1988), "Subgame Perfect Implementation," *Econometrica* 58: 1083–1100.

Mutuswami, S., and E. Winter (2002), "Subscription Mechanisms for Network Formation," *Journal of Economic Theory* 106: 242–264.

Index

Printed in the United States
by Baker & Taylor Publisher Services